PRINCIPLES OF BIOSEPARATIONS ENGINEERING

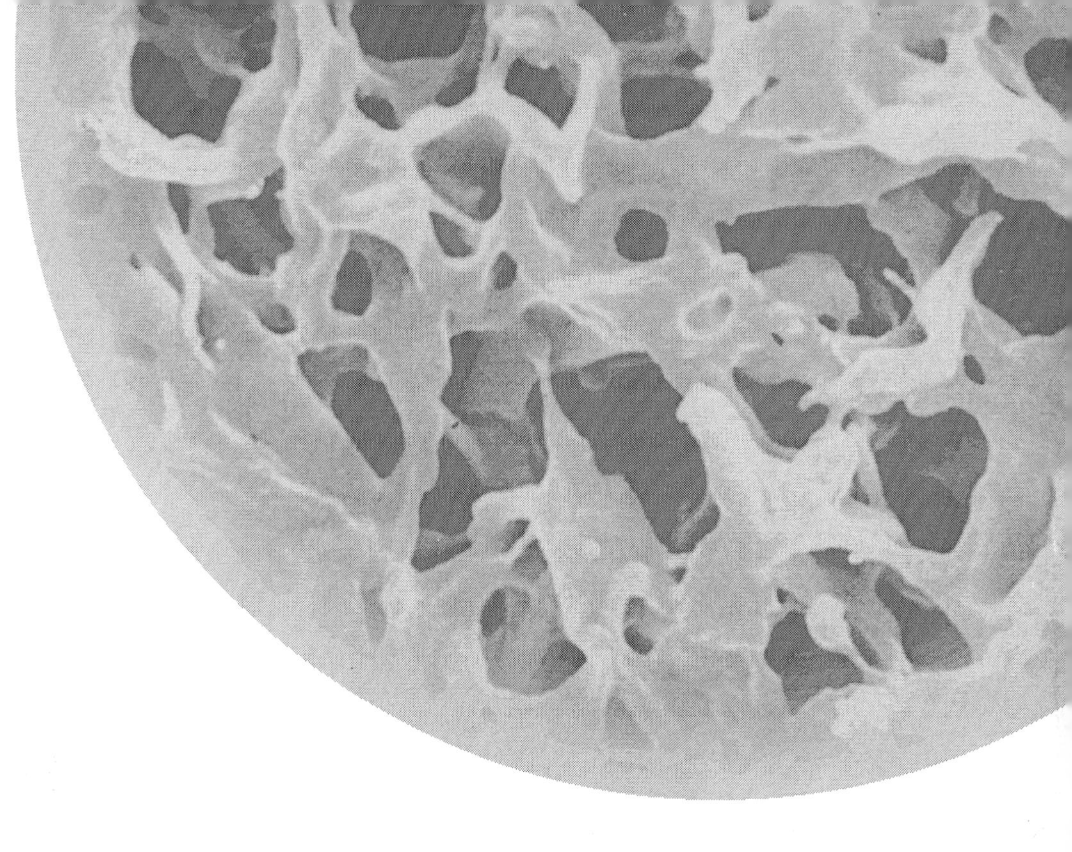

PRINCIPLES OF BIOSEPARATIONS ENGINEERING

RAJA GHOSH
McMaster University, Canada

NEW JERSEY · LONDON · SINGAPORE · BEIJING · SHANGHAI · HONG KONG · TAIPEI · CHENNAI

Published by

World Scientific Publishing Co. Pte. Ltd.
5 Toh Tuck Link, Singapore 596224
USA office: 27 Warren Street, Suite 401-402, Hackensack, NJ 07601
UK office: 57 Shelton Street, Covent Garden, London WC2H 9HE

British Library Cataloguing-in-Publication Data
A catalogue record for this book is available from the British Library.

Cover: Image courtesy of Millipore Corporation.

PRINCIPLES OF BIOSEPARATIONS ENGINEERING

ISBN 981-256-892-1

Printed in Singapore by World Scientific Printers (S) Pte Ltd

To my wife, Sutapa

Acknowledgements

The author acknowledges with thanks the assistance provided by the individuals and organisations named below:

Christine Zardecki of RCSB Protein Data Bank for kind permission to use images of protein and DNA structures (Figures 2.1, 2.2 and 2.3 of this book)

Stanley Goldberg and Marc Lustig of Glen Mills Inc. and Andre Wrysch W. A. of Bachofen AG for providing me with Figures 4.5, 4.10 and 4.11 of this book and for their enthusiastic support for this project

Dominic Spricigo of IKA® -WERKE GmbH & Co. KG for providing me with Figure 4.7 of this book

Kevin McLaughlin of Shimadzu Scientific Instruments for providing me with Figure 9.4 of this book

Gene Lauffer of Menardi for kind permission to use images of filtration devices from their website (Figures 10.7 and 10.10 of this book)

Dana Hubbard, Susan Cheah and Shannon Meirzon of Millipore Corporation for providing me with Figures 11.4 (A, B and C), 11.12, 11.13 and 11.14 of this book

PCI Membranes for providing me with Figure 11.15 of this book

Shuichi Ogawa of Asahi Kasei for kind permission to use image of membrane filtration device from their website (Figure 11.16 of this book)

Patricia Stancati of Sartorius for providing me with Figure 11.39 of this book

Sandra Covelli and Michael Killeen of Pall Corporation for providing me with Figure 11.41 of this book

Preface

Bioseparations engineering refers to the systematic study of the scientific and engineering principles utilized for large-scale purification of biological products: biopharmaceuticals, biochemicals, foods, nutraceuticals and diagnostic reagents. Bioseparations engineering, both as an academic topic as well as an industrial practice has undergone significant growth and increase in importance in the past decade. Its entry into academia as a taught course took place as usually happens with any new topic: at the postgraduate level. It is now taught as a regular course in many undergraduate chemical and biomolecular engineering programs. It is also taught either as a separate course or indeed as a component of *bioprocess engineering* in almost all undergraduate and postgraduate programs in biotechnology.

Most projects start with some necessity and this book is no exception. In the year 1999, as a new faculty member at Oxford University I was assigned to teach the relatively new *C5A Bioprocess Engineering* and *C5B Separations* courses for undergraduate engineering students. Both these courses were expected to have significant *bioseparations engineering* content and the main challenge I faced at that time was the lack of a satisfactory undergraduate level text book. I had learnt *bioseparations engineering* in the early 90s reading the seminal book on this topic by Belter, Cussler and Hu (*Bioseparations: Downstream Processing for Biotechnology, John Wiley and Sons*). However, this book had been published in 1988 and by end of the millennium was slightly dated. I therefore started developing my own course notes and handouts. These have formed the building blocks of this book and so in a sense, I effectively started writing this book in 1999. This is intended to be an undergraduate level text book, suitable for a one semester course on bioseparations engineering.

Chapters 1 to 3 are intended to provide the reader with the basic biological and engineering prerequisites. Most bioseparation processes are carried out in a multi-step fashion, each step consisting of a discrete separation technique. These techniques have been classified into broad categories in the remaining chapters of the book, i.e. chapters 4 to 12.

The last paragraph of a preface is usually reserved for acknowledgements. I would like to thank my family members: my parents, my wife, my sister and parents in-law for their encouragement and unstinting support. My wife has been a constant source of moral and material support throughout this project. I am particularly grateful to her for painstakingly proof-reading the manuscript, making it better in the process. Without her help, it would have been very difficult to complete this book. I would like to acknowledge my McMaster colleagues Professors Carlos Filipe, Ronald Childs, Andrew Hrymak, Robert Pelton, Shiping Zhu and Don Woods for kind words of encouragement and support. Going international, I would like to thank Professor Tony Fane of the University of New South Wales, Australia and Professor Zhanfeng Cui of the University of Oxford, UK for their words of advice at the early stages of this project. Last but certainly not the least: I would like to acknowledge my graduate students Dharmesh Kanani and Lu Wang for their help with crosschecking the solutions to the numerical problems presented in this book.

Raja Ghosh
Hamilton, ON
April 2006

Contents

Chapter 1

Overview of bioseparations engineering

1.1. Introduction

Biotechnology has undergone phenomenal growth in recent years. Most of this has been in fundamental areas linked to cellular and molecular biology, biochemistry and biophysics. In comparison, growth in more applied areas such as *bioprocess engineering* has been relatively modest, largely due to significantly lower funding in these areas, both in academia and industry. The net result of this has been the hugely inflated but largely unfulfilled expectations about the benefits of biotechnology. It is now being appreciated that symmetric growth in basic and applied areas is crucial for healthy development of biotechnology. One of the major segments within biotechnology where research and development is vital is *bioprocessing* which deals with the manufacture of biochemicals, biopharmaceuticals, foods, nutraceuticals, and agrochemicals. A plethora of new biologically derived products have been developed, approved and licensed in the last decade. This includes monoclonal antibodies used for the treatment of cancer and multiple sclerosis, plasmids for gene therapy, cytokines and interleukins. Many of these products need to be extensively purified before they can be used for their respective applications. *Bioseparations engineering* refers to the systematic study of the scientific and engineering principles utilized for the large-scale purification of biological products. It is a broader term than the slightly dated *downstream processing* which specifically referred to the separation and purification segment of a bioprocess which followed some form of biological reaction e.g. purification of an antibiotic following microbial fermentation. However, the manufacture of several types of biological products does not involve *in-vitro* biological reactions. These products are synthesized *in vivo* in their respective natural sources and these are simply recovered using appropriate techniques e.g. manufacture of plasma proteins from blood, extraction of alkaloids from plants, extraction of enzymes from animal tissue. Bioprocessing can be broadly classified into two categories (see Fig. 1.1):

1. Reactive bioprocessing
2. Extractive bioprocessing

In reactive bioprocessing, the bioseparation process follows some form of biological reaction whereas extractive bioprocessing almost entirely involves bioseparation. In the context of reactive bioprocessing, upstream processing involves steps such as biocatalyst screening, enrichment, isolation and propagation, cell manipulation by recombinant DNA technology or hybridoma technology, media optimization and formulation, and so on. The biological reaction involved could be fermentation (i.e. cultivation of bacterial or fungal cells), cell culture (i.e. cultivation of animal or plant cells) or simply an enzymatic reaction. With extractive bioseparation, upstream processing involves raw material acquisition and pre-treatment.

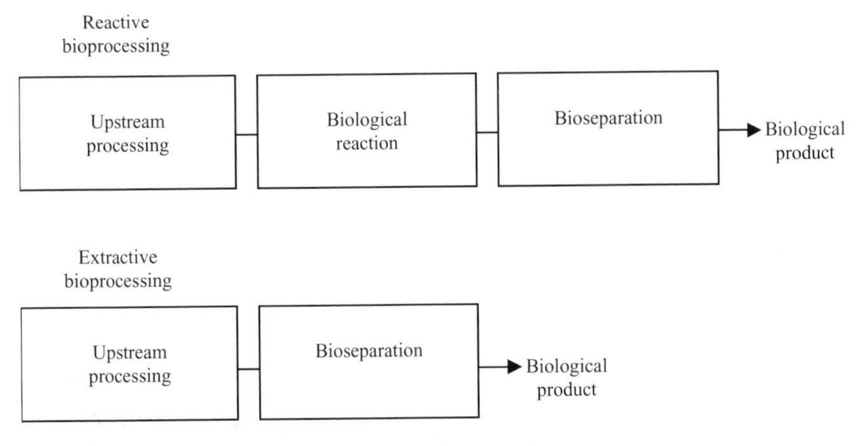

Fig. 1.1 Bioprocessing

1.2. What is separated in bioseparation?

Biologically derived products can be categorized in different ways, one way being based on their chemical nature (see Table 1.1).

Table 1.1 Biological products classified based on chemical nature

Categories	**Examples**
Solvents	Ethanol, acetone, butanol
Organics acids	Citric acid, lactic acid, butyric acid
Vitamins	Ascorbic acid, vitamin B12
Amino acids	Lysine, phenylalanine, glycine
Antibiotics	Penicillins, rifanpicin, streptomycin, framycetin, gentamycin
Sugars and carbohydrates	Glucose, fructose, starch, dextran, xanthan, gellan
Lipids	Glycerol, fatty acids, corticosteroids, prostaglandins
Nucleic acids	Plasmids, therapeutic DNA, retroviral vectors, anti-sense oligonucleotides, ribozymes
Semi-purified proteins	Industrial enzymes, egg proteins, milk proteins, whey protein, soy proteins
Purified proteins	Therapeutic enzymes, monoclonal antibodies, plasma proteins, cytokines, interleukins, hormones, growth factors, diagnostic enzymes, vaccines
Cells	Bakers yeast, brewers yeast, freeze dried lactobacillus
Crude cellular extracts	Yeast extract, soy extracts, animal tissue extract
Hydrolysates	Soy hydrolysates, whey hydrolysates, animal tissue hydrolysates

Biological products can also be classified based on their intended applications (Table 1.2):

Table 1.2 Biological products classified based on application

Categories	**Examples**
Industrial chemicals	Solvents, organic acids, industrial enzymes
Agrochemicals	Biofertilizers, biopesticides
Biopharmaceuticals	Antibiotics, hormones, monoclonal antibodies, plasma proteins, vaccines, hormones, cytokines, therapeutic nucleic acids
Food and food additives	Whey proteins, milk proteins, egg proteins, soy proteins, protein hydrolysates
Nutraceuticals	Vitamins, enzymes, coenzymes, cofactors, amino acids, purified whey proteins
Diagnostic products	Glucose oxidase, peroxidase, HCG
Commodity chemicals	Detergent enzymes, insecticides
Laboratory reagents	Bovine serum albumin, ovalbumin, lysozyme
Cosmetic products	plant extracts, animal tissue extracts

1.3. Economic importance of bioseparation

The purification of biological products from their respective starting material e.g. cell culture media is technically difficult and expensive. This could frequently be the critical limiting factor in the commercialization of a biological product. In many cases bioseparation cost can be a substantial component of the total cost of bioprocessing. Table 1.3 summarizes the bioseparation cost of different categories of biological products. For proteins and nucleic acids, particularly those used as biopharmaceuticals, the bioseparation cost is quite substantial.

Table 1.3 Cost of bioseparation

Products	Bioseparation cost (%)
Solvents	15-20
Cells	20-25
Crude cellular extracts	20-25
Organics acids	30-40
Vitamins and amino acids	30-40
Gums and polymers	40-50
Antibiotics	20-60
Industrial enzymes	40-65
Non-recombinant therapeutic proteins	50-70
r-DNA products	60-80
Monoclonal antibodies	50-70
Nucleic acid based products	60-80
Plasma proteins	70-80

1.4. Nature of bioseparation

Bioseparation is largely based on chemical separation processes. A plethora of well established separation techniques is used in the chemical industry. A number of these techniques were found to be suitable for carrying out biological separations. However, while borrowing from chemical separations, the fundamental differences between synthetic chemicals and biological substances need to be kept in mind. Some biologically derived substance such as antibiotics and other low molecular weight compounds such as vitamins and amino acids are purified using conventional separation techniques such as liquid-liquid extraction, packed bed adsorption, evaporation and drying with practically no modifications being necessary. However, substantially modified separation techniques are required for purifying more complex molecules such as proteins, lipids, carbohydrates and nucleic acids. Often totally new types of separation techniques have to be devised. Some of the attributes of bioseparation which distinguish it from chemical separation are:

1. Biological products are present in very low concentrations in the starting material from which they are purified. For example, monoclonal antibodies are typically present in concentrations around 0.1 mg/ml in the mammalian cell culture supernatants. Hence large volumes of dilute product streams have to be processed for obtaining even modest amounts of pure products.
2. Several other substances which are usually impurities and in some instances by-products are present in the starting material along with target biological products. Frequently these impurities or by-products have chemical and physical properties similar to those of the target product. This makes separation extremely challenging. Hence, bioseparation has to be very selective in nature.
3. There are stringent quality requirements for products used for prophylactic, diagnostic and therapeutic purposes both in terms of active product content as well as in terms of the absence of specific impurities. Injectable therapeutic products should be free from endotoxins and pyrogens. Solutions for such specific requirements have to be built into a bioseparation process.
4. Biological products are susceptible to denaturation and other forms of degradation. Therefore bioseparation techniques have to be "gentle" in terms of avoiding extremes of physicochemical conditions such as pH and ionic strengths, hydrodynamic conditions such as high shear rates, and exposure to gas-liquid interfaces. Organic solvents which are widely used in chemical separations have relatively limited usage in bioseparations on account of their tendency to promote degradation of many biological products.
5. Many biological products are thermolabile and hence many bioseparation techniques are usually carried out at sub-ambient temperatures.
6. Bioseparation is frequently based on multi-technique separation. This will be discussed in detail in a subsequent section.

1.5. Basis of separation in bioseparation processes

Biological products are separated based on one or more of the following factors:

1. Size: e.g. filtration, membrane separation, centrifugation

2. Density: e.g. centrifugation, sedimentation, floatation
3. Diffusivity: e.g. membrane separation
4. Shape: e.g. centrifugation, filtration, sedimentation
5. Polarity: e.g. extraction, chromatography, adsorption
6. Solubility: e.g. extraction, precipitation, crystallization
7. Electrostatic charge: e.g. adsorption, membrane separation, electrophoresis
8. Volatility: e.g. distillation, membrane distillation, pervaporation

1.6. Physical forms separated in bioseparation

Bioseparation usually involves the separation of the following physical forms:

1.6.1. Particle-liquid separation

Examples of particle-liquid separation include the separation of cells from cell culture medium, the separation of blood cells from plasma in the manufacture of plasma proteins, and the removal of bacteria and viruses from protein solutions. Particle-liquid separation can be achieved by forcing the suspension through a porous medium which retains the particles while allowing the liquid to go through. This principle is utilized in filtration and membrane separation. Particle-liquid separation can also be achieved by subjecting the suspension to natural or artificially induced gravitational fields. If the particles are denser than the liquid medium in which they are suspended, these would settle and form a zone with very high particulate concentration. This is referred to as the sediment and the clear liquid left behind is referred to as the supernatant. This principle is utilized in separation processes such as sedimentation and centrifugation. If the particles are lighter than the liquid in which they are suspended, these would tend to float and hence concentrate near the top of the container in which the suspension is held. This principle is utilized in floatation.

1.6.2. Particle-particle separation in liquid medium

Examples of particle-particle separation in liquid medium include the fractionation of sub-cellular organelle, the separation of plasmid DNA from chromosomal DNA, and the separation of mature cells from young cells. This type of separation can be achieved by zonal centrifugation which involves the introduction of the mixture at a location within a liquid medium which is then subjected to an artificially induced gravitational field. As a result of this the heavier particles would migrate faster than the lighter particles, resulting in their segregation into distinct bands from which these particles can be subsequently recovered using appropriate means. Particle-particle separation can in theory be carried out by using a porous medium which retains the bigger particles but allows the smaller particles to go through. However, this sounds easier than it actually is and can only be carried out if the larger particles can be prevented from blocking the porous medium.

1.6.3. Particle-solute separation in liquid medium

An example of this is the separation of dissolved antibiotics from cells and cell debris present in fermentation broth. The methods used for particle-solute separation are fundamentally similar to those used for solid-liquid separation on account of the fact that the solute remains dissolved in the liquid medium.

1.6.4. Solute-solvent separation

Solute-solvent separation is quite common in bioseparation, the purpose of this being either the total or partial removal of a solvent from a solute product (e.g. protein concentration enrichment), or the removal of dissolved impurities from a liquid product, or the replacement of a solvent from a solution by another (i.e. solvent exchange). A range of options are available for solute-solvent separation the easiest of these being evaporation and distillation. However, these techniques involve the application of heat and cannot therefore be used for separation of biological materials which tend to be thermolabile. Membranes which can retain dissolved material while allowing solvents through are widely used for this type of separation: a reverse osmosis membrane will retain small molecules and ions, a nanofiltration membrane will retain larger molecules such as vitamins, hormones and antibiotics, while an ultrafiltration membrane will retain macromolecules such as proteins and nucleic acids. Another way of removing a solvent from a solute is by reversibly binding the solute on to a solid surface, this being referred to as adsorption. Once solute binding has taken place, this separation is transformed to a particle-liquid separation, i.e. the solvent is separated from the solid-bound solute. The bound solute is subsequently recovered from the solid material, this being referred to as desorption. An indirect method for solute-solvent separation is by inducing precipitation of the solute, thereby once again transforming the separation to a particle-liquid separation. Solvent exchange can also be carried out by liquid-liquid extraction where the solute is transferred from a liquid to another with which the original solvent is immiscible.

1.6.5. Solute-solute separation in liquid medium

Solute-solute separation is by far the most challenging form of separation. An example of this is the separation of serum albumin from other serum proteins. Solute-solute separation can be achieved by selective adsorption, i.e. by selectively and reversibly binding the target solute on to a solid material. Solute-solute separation can also be carried out by liquid-liquid extraction, i.e. by contacting the solution with an immiscible liquid in which the target solute has high solubility. With the advent of membranes, solute-solute separation has become a lot easier. Nanofiltration, ultrafiltration and dialysis membranes can be used for such separations. An indirect way of carrying out solute-solute separation is by precipitation, which involves the selective precipitation of the target solute. This separation is then transformed to that of particle-solute separation in liquid medium.

1.6.6. Liquid-liquid separation

Liquid-liquid separation is required in the manufacture of solvents such as acetone and ethanol which typically have to be separated from an aqueous medium. If the solvent is immiscible with water, phase separation followed by decantation may be sufficient. However, if the solvent is miscible with water (as in the case of ethanol), other separation methods have to be utilized. With temperature stable and volatile solvents such as ethanol, distillation has been traditionally used. However with the advent of membrane technology, separation processes such as membrane distillation and pervaporation have come into widespread use.

1.7. Bioseparation techniques

A plethora of bioseparation techniques is now available. Table 1.4 categorizes bioseparation techniques into two broad groups. As previously mentioned, a bioseparation process must combine high selectivity (or resolution) with high throughput (or productivity). Quite clearly none of those listed in the table can deliver this on their own.

Hence bioseparation processes tend to be based on multiple techniques arranged such that both high-resolution and high-throughput can be obtained in an overall sense.

Table 1.4 Bioseparation techniques

Low-resolution + high-throughput	**High-resolution + low-throughput**
Cell disruption	Ultracentrifugation
Precipitation	Chromatography
Centrifugation	Affinity separation
Liquid-liquid extraction	Electrophoresis
Leaching	
Filtration	
Supercritical fluid extraction	
Microfiltration	
Ultrafiltration	
Adsorption	

1.8. The RIPP scheme

While developing a bioseparation process the following should be taken into consideration:

1. The nature of starting material: e.g. a cell suspension, a crude protein solution
2. The initial location of the target product: e.g. intracellular, extracellular, embedded in solid material such as inclusion bodies
3. The volume or flow-rate of the starting material
4. The relative abundance of the product in the starting material, i.e. its concentration relative to impurities
5. The susceptibility to degradation e.g. its pH stability, sensitivity to high shear rates or exposure to organic solvents
6. The desired physical form of the final product, e.g. lyophilized powder, sterile solution, suspension
7. The quality requirements, e.g. percentage purity, absence of endotoxins or aggregates
8. Process costing and economics

A RIPP (Recovery, Isolation, Purification and Polishing) scheme is commonly used in bioseparation. This strategy involves use of low-resolution techniques (e.g. precipitation, filtration, centrifugation, and crystallization) first for recovery and isolation followed by high-resolution techniques (e.g. affinity separations, chromatography, and electrophoresis) for purification and polishing. The high-throughput, low-resolution techniques are first used to significantly reduce the volume and overall concentration of the material being processed. The partially purified products are then further processed by high-resolution low-throughput techniques to obtain pure and polished finished products.

1.9. Example of bioseparation

A scheme for the bioseparation of reagent grade monoclonal antibody from cell culture supernatant is shown in Fig. 1.2. Murine or mouse monoclonal antibodies are produced by culturing hybridoma cells in different types of bioreactors. In recent years it has been possible to synthesize humanized and chimaeric monoclonal antibodies by culturing recombinant Chinese Hamster Ovarian (CHO) cells. In the bioseparation scheme shown in Fig. 1.2, the key purification step involves affinity chromatography. Prior to affinity chromatography the cell culture supernatant needs to be cleaned up by membrane filtration or centrifugation so that cells, cell debris and other particulate matter do not clog-up the affinity column. The nearly purified monoclonal antibody obtained by affinity chromatography is further purified by ion-exchange chromatography and polished by gel-filtration to obtain greater than 98% pure product in the solution form. This percentage purity figure is relative to other proteins present in the product. The antibody solution is then filtered to remove bacterial contaminant and marketed either as a sterile solution or as a freeze dried powder. The scheme for purifying therapeutic grade monoclonal antibodies would be largely similar to that shown in Fig. 1.2. In addition to the basic purification scheme used for making the reagent grade monoclonal antibody, some additional steps for removing particulate matter and specific impurities such as endotoxins and antibody dimers and higher order aggregates would be required. An additional step to formulate the monoclonal antibody in an appropriate buffer would also be required.

1.10. Current trends in the bioseparation

The main disadvantages of using the RIPP scheme are:

1. High capital cost
2. High operations cost
3. Lower recovery of product

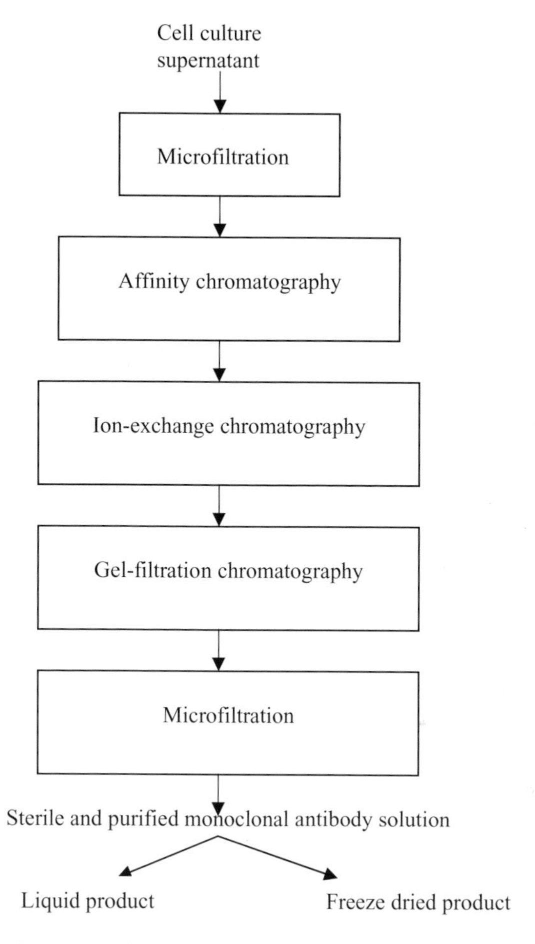

Fig. 1.2 Purification of reagent grade monoclonal antibody

With the advent of membrane separation processes and other new types of separations, the potential exists for avoiding the conventional RIPP scheme. Membrane processes give high throughput and can be fine-tuned or optimized to give very high selectivity. The use of these new techniques can significantly cut down the number of steps needed for bioseparation. Some of these new and emerging techniques are:

1. Membrane and monolith chromatography
2. Expanded-bed chromatography
3. High-resolution ultrafiltration
4. Hybrid bioseparations

References

P.A. Belter, E.L. Cussler, W.-S. Hu, Bioseparations: Downstream Processing for Biotechnology, John Wiley and Sons, New York (1988).

R. Ghosh, Protein Bioseparation Using Ultrafiltration: Theory Applications and New Developments, Imperial College Press, London (2003).

M.R. Ladisch, Bioseparations Engineering: Principles, Practice and Economics, John Wiley and Sons, New York (2001).

P. Todd, S.R. Rudge, D.P. Petrides, R.G. Harrison, Bioseparations Science and Engineering, Oxford University Press, Oxford (2002).

G. Walsh, Biopharmaceuticals: Biochemistry and Biotechnology, John Wiley and Sons, New York (2003).

Chapter 2

Properties of biological material

2.1. Introduction

Before discussing how bioseparation is carried out it is perhaps useful to discuss some fundamental properties of biological substances, particularly those that are relevant in separation processes:

1. Size
2. Molecular weight
3. Diffusivity
4. Sedimentation coefficient
5. Osmotic pressure
6. Electrostatic charge
7. Solubility
8. Partition coefficient
9. Light absorption
10. Fluorescence

2.2. Size

Table 2.1 lists some biological substances with their respective sizes. For macromolecules such as proteins and nucleic acids the molecular size cannot always be represented in terms of a quantity such as the diameter, particularly when these molecules are non-spherical in shape. For ellipsoid molecules (see Fig. 2.1) or for those where a major and a minor axis can be identified, size is frequently shown in terms of a "length" i.e. dimension along the major axis and a "breadth", i.e. the dimension along the minor axis. Some molecules such as antibodies have even more complex shapes (see Fig. 2.2). A universal way to express the dimension of non-spherical species is in terms of the Stokes-Einstein diameter. This is the diameter of a spherical molecule or particle having the same diffusivity, i.e. a protein molecule having a Stokes-Einstein diameter of 10 nanometers (nm) has the same diffusivity as a sphere of same density having a diameter of 10 nm. Nucleic acids such as DNA and RNA are linear molecules (see Fig. 2.3) and their sizes cannot be expressed in terms of the Stokes-Einstein diameter. Instead their sizes are expressed in terms of their lengths alone.

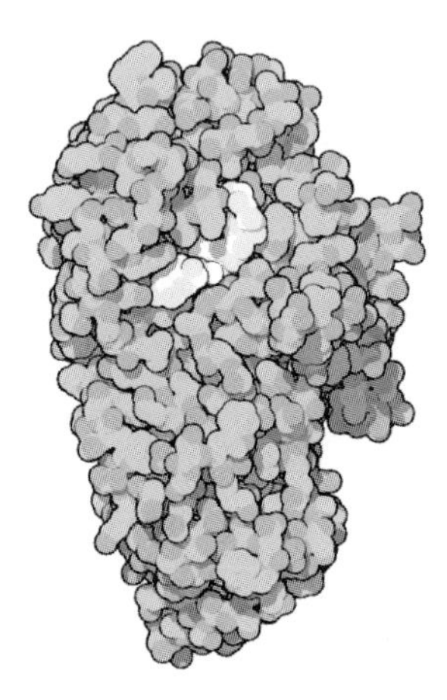

Fig. 2.1 Alpha amylase, an example of an ellipsoid molecule (Image courtesy of RCSB Protein Data Bank)

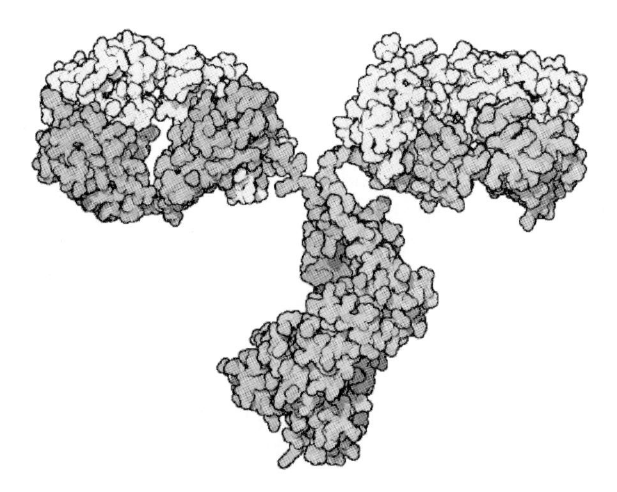

Fig. 2.2 Structure of an antibody molecule (Image courtesy of RCSB Protein Data Bank)

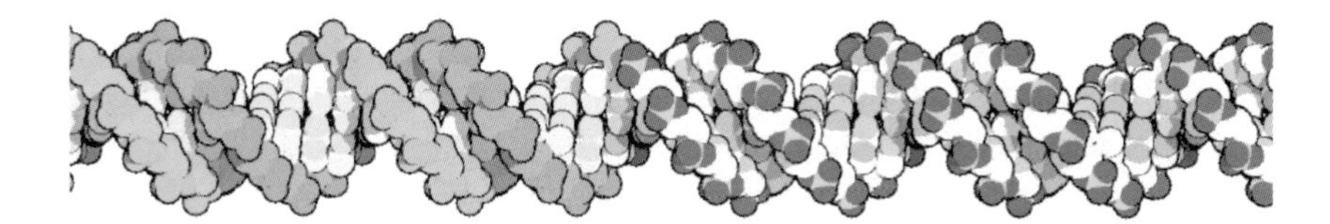

Fig. 2.3 Structure of DNA molecule (Image courtesy of RCSB Protein Data Bank)

Table 2.1 Size of biological material

Material	**Size (microns)**
Chloride ion	9.9×10^{-5}
Urea	1.6×10^{-4}
Glucose	8.6×10^{-4}
Insulin	4×10^{-3}
Human serum albumin	7.2×10^{-3}
Dextran	1×10^{-3} to 15×10^{-3}
Immunoglobulin G	11.1×10^{-3}
Virus particle	0.004 to 0.1
Bacteria	0.2 to 2.0
Yeast cells	2 to 30
Mammalian cell in culture	2-10

The size of biological material is important in separation processes such as conventional filtration, membrane separation, sedimentation, centrifugation, size exclusion chromatography, gel electrophoresis, hydrodynamic chromatography, to name just a few. The size of particulate matter such as cells, cell debris and macromolecular aggregates can be measured by direct experimental techniques such optical and electron microscopy. Indirect methods such as the Coulter counter technique or laser light scattering techniques are also used for determining particle size. For dense particles, the sedimentation rate, i.e. the rate of settling under gravity in a fluid having a lower density can be used to measure particle size. Gravitational settling is feasible only with particles larger than 5 microns in diameter. The equivalent radius (r_e) of a particle settling under gravity can be estimated from its terminal velocity:

$$r_e = \sqrt{\frac{9\mu u_T}{2g(\rho_s - \rho_l)}} \tag{2.1}$$

Where

μ = viscosity (kg/m s)
u_T = terminal velocity (m/s)
ρ_s = density of the particle (kg/m^3)
ρ_l = density of the liquid medium (kg/m^3)

Example 2.1
A suspension of kaolin (a type of clay used as adsorbent for biological material) in water became clear upon being allowed to stand undisturbed for 3 minutes at 20 degrees centigrade. The height of the suspension in the vessel was 30 cm and the density of kaolin is known to be 2.6 g/cm^3. Estimate the diameter of the kaolin particles.

Solution
In this problem, we have to make the following assumptions:

1. Complete clarification coincided with the movement of the particles from the topmost portion of the suspension to the bottom of the vessel.
2. The terminal velocity of the settling kaolin particles is quickly reached such that the particles settle uniformly at this velocity throughout their settling distance.

The density of water at 20 degrees centigrade is 1 g/cm^3 while its viscosity is 1 centipoise (= 0.01 poise). The terminal velocity of the particles is (30/180) cm/s = 0.167 cm/s. The acceleration due to gravity is 981 cm/s^2.
Using equation (2.1):

$$r_e = \sqrt{\frac{9 \times 0.01 \times 0.167}{2 \times 981 \times (2.6 - 1)}} \text{ cm} = 2.19 \times 10^{-3} \text{ cm}$$

Therefore the diameter is 4.38×10^{-3} cm.

Microbial, animal or plant cells in a given sample are usually not all of the same size due to the different levels of growth and maturity in a given population, i.e. these demonstrate various particle size distributions. For such particulate systems which are referred to as polydispersed systems, the representative particle size is expressed in terms of statistically determined values such as the average diameter or the median diameter. The most common form of particle size distribution is the normal or Gaussian distribution which has one mode, i.e. is mono-modal. In some cases, cell suspensions can show bi-modal distribution.

With macromolecules such as proteins and nucleic acids, the size can be estimated using indirect methods. The classical laser light scattering technique works reasonably well with larger macromolecules and macromolecular aggregates such as aggregated antibodies. The sample is held in a chamber and laser light is shown on it. The angle at which an incident light is scattered by these substances depends on their size and hence by measuring light at different angles, inferences about size and size distribution can be made. However, most smaller and medium sized proteins cannot be satisfactorily resolved by classical laser scattering techniques on account of the fact that these scatter light uniformly in all directions. Dynamic laser light scattering technique which measures subtle variations in light scattering at different locations within a sample can give valuable information about mobility of molecules from which hydrodynamic dimensions can be estimated. Other indirect methods such as hydrodynamic chromatography, size-exclusion chromatography, and indeed diffusion and ultracentrifugation based techniques can be used to measure the size of macromolecules and particles. The Stokes-Einstein radius (r_{SE}) of a macromolecule can be estimate from its diffusivity using the following equation:

$$r_{SE} = \frac{RT}{6\pi D \mu N} \qquad (2.2)$$

Where

D = diffusivity (m^2/s)

μ = viscosity (kg/m s)

N = Avogadro's number

2.3. Molecular weight

For macromolecules and smaller molecules, the molecular weight is often used as a measure of size. Molecular weight is typically expressed in Daltons (Da) or g/g-mole or kg/kg-mole. Table 2.2 lists the molecular weights of some biological substances. With nucleic acids, such as plasmids and chromosomal DNA, the molecular weight is frequently expressed in term of the number of base pairs of nucleotides present (bp). One base pair is roughly equivalent to 660 kg/kg-mole.

Molecular weight being linked to size is used as a basis for separation in techniques such as gel-filtration, hydrodynamic chromatography and membrane separations. The molecular weight of a substance also influences other properties of the material such as sedimentation, diffusivity and mobility in an electric field and can hence be an indirect basis for separation in processes such as ultracentrifugation and electrophoresis.

Table 2.2 Molecular weights of biological material

Material	Molecular weight (kg/kg-mole)
Sodium chloride	58.44
Urea	60.07
Glucose	180
Penicillin G	334.42
Insulin	5,800
Lysozyme	14,100
Myoglobin	17,000
Human serum albumin	67,000
Immunoglobulin G	155,000
Dextrans	10,000 – 1000,000
Catalase	250,000
Collagen	345,000
Myosin	493,000
Tobacco mosaic virus	40,590,000
Plasmids	Few kilo-base pairs
Chromosomal DNA	Hundreds of kilo-base pairs

As with particle size, the molecular weight of certain substances can be polydispersed i.e. may demonstrate a molecular weight distribution. Examples include the polysaccharides dextran and starch, both of which have very large molecular weight ranges. A special case of a polydispersed system is that of a paucidispersed system, where molecular weights in a distribution are multiples of the smallest molecular weight in the system. An example of paucidispersed system is immunoglobulin G in solution which occurs predominantly as the monomer with presence of dimers and smaller amounts of trimers and tetramers. The molecular weight of small molecules can easily be determined based on their structural formula. Molecular weights of macromolecules are usually determined using experimental methods such as hydrodynamic chromatography, size-exclusion chromatography and ultracentrifugation.

Size-exclusion chromatography is a column based method where separation takes place based on size. If a pulse of sample containing molecules of different molecular weights is injected into one end of a size-exclusion column, the larger molecules appear at the other end of the column earlier than the smaller ones. The molecular weights of known sample can be calibrated against their corresponding exit times and based on this the molecular weights of unknown samples can be estimated. This technique is discussed in the chapter on *chromatography*. Hydrodynamic chromatography shows a similar behaviour, i.e. size-based segregation, but the exact mechanism determining exit time is different from that in size-exclusion chromatography. This technique is also discussed in the chapter on *chromatography*.

2.4. Diffusivity

Diffusion refers to the random motion of molecules due to inter-molecular collision. Even though the collisions between molecules are random in nature, the net migration of molecules takes place from a high concentration to a low concentration zone. The diffusivity or diffusion coefficient is a measure of the molecules tendency to diffuse, i.e. the greater the diffusion coefficient, the greater is its mobility in response to a concentration differential. Diffusivity is an important parameter in most bioseparation processes since it affects material transport. Table 2.3 lists the diffusivities of some biological substances. The diffusion coefficient can be measured using some of the experimental techniques discussed in the previous section. Some specific methods for measuring diffusivity will be discussed in the next chapter. Diffusivity is primarily dependent on the molecular weight but is also influenced by the friction factor of the molecule and the viscosity of the medium. The friction factor depends on the shape of the molecule as well as on the degree of hydration (if the molecule is present in an aqueous system). The diffusivity of a molecule correlates with its Stokes-Einstein radius as shown in equation 2.2. The manner in which diffusivity influences the transport of molecules is discussed in the next chapter.

Table 2.3 Diffusivity of biological material

Material	**Diffusivity in water (m^2/s)**
Sodium chloride	2.5×10^{-9}
Urea	1.35×10^{-9}
Acetic acid	1.19×10^{-9}
Ethanol	8.4×10^{-10}
Glycerol	7.2×10^{-10}
Lysozyme	10.4×10^{-11}
Human serum albumin	5.94×10^{-11}
Human immunoglobulin G	4.3×10^{-11}
Collagen	0.69×10^{-11}
Tobacco mosaic virus	0.46×10^{-11}

Example 2.2
The diffusivity of a protein having a Stokes-Einstein radius of 2 nanometers in a particular liquid is known to be 4.5×10^{-11} m^2/s. Predict the diffusivity of another molecule having twice the Stokes-Einstein radius in the same liquid at the same temperature.

Solution
Using equation (2.2):

$$\frac{r_{SE\,1}}{r_{SE\,2}} = \frac{D_2}{D_1}$$

Where 1 and 2 stand for the two molecules under consideration

Therefore:

$$D_2 = \frac{4.5 \times 10^{-11}}{2} \text{ m}^2/\text{s} = 2.25 \times 10^{-11} \text{ m}^2/\text{s}$$

2.5. Sedimentation coefficient

The tendency of macromolecules and particles to settle in a liquid medium is referred to as sedimentation. This is the basis of separation in processes such as decantation, centrifugation and ultracentrifugation. Settling could take place due to gravity as in decantation or due to an artificially induced gravitational field as in centrifugation where this field is generated by rotating about an axis, a vessel containing the particles or macromolecules dispersed in a liquid medium. The rate of settling depends on the properties of the settling species as well as those of the liquid medium. These include their respective densities and the frictional factor. The rate of settling also depends on the strength of the gravitational field which in centrifugation depends on the geometry of the vessel, the location within the vessel and on the speed at which the vessel is rotated. The sedimentation coefficient (s) of a particle/macromolecule in a liquid medium can be expressed in terms of operating parameters of the centrifugation process as shown below:

$$s = \frac{v}{\omega^2 r} \tag{2.3}$$

Where

v = sedimentation velocity (m/s)

ω = angular velocity of rotation (radians/s)

r = distance from the axis of rotation (m)

The sedimentation coefficient correlates with the material properties as shown below:

$$s = \frac{M(1 - v_M \rho)}{f} \tag{2.4}$$

Where

M = molecular weight (kg/kg-mole)
v_M = partial specific molar volume (m^3/kg)
ρ = density (kg/m^3)
f = frictional factor

The sedimentation coefficients of some biological substances in c.g.s. units are listed in Table 2.4. The subscript 20 indicates that these values were obtained at 20 degrees centigrade.

Table 2.4 Sedimentation coefficients of biological material

Material	**Sedimentation coefficient** s_{20}
Cytochrome c	1.17
Chymotrypsinogen	2.54
Alpha amylase	4.5
Fibrinogen	7.9
Human serum albumin	4.6
Collagen	6.43
Tobacco mosaic virus	198

2.6. Osmotic pressure

If a dilute aqueous solution of any solute is separated from a concentrated one by a semi-permeable membrane that only allows the passage of water, a pressure differential is generated across the membrane due to the tendency of water to flow from the low to high solute concentration side. This is referred to as osmotic pressure. This concept was first described by the French physicist Jean-Antoine Nollet in the 18th century. Osmotic pressure has a significant role in bioseparations, particularly in membrane based separation processes. The osmotic pressure can be correlated to the solute concentration. For dilute solutions, the van't Hoff equation can be used to estimate osmotic pressure (π):

$$\pi = RTc \tag{2.5}$$

Where

R = universal gas constant
T = absolute temperature (K)
c = solute concentration (kg-moles/m^3)

For concentrated solutions of uncharged solutes, correlations involving series of virial coefficients are used:

$$\pi = RT\left(A_1 C + A_2 C^2 + A_3 C^3 + ...\right) \quad (2.6)$$

Where
A_1 = constant which depends on the molecular weight
A_2 = second virial coefficient
A_3 = third virial coefficients
C = solute concentration (kg/m^3)

The osmotic pressure difference across a membrane is given by:

$$\Delta\pi = \pi_1 - \pi_2 \quad (2.7)$$

Where
1 represents the higher concentration side
2 represents the lower concentration side

It should be noted that the osmotic pressure acts from the lower concentration side to the higher concentration side.

2.7. Electrostatic charge

Ions such as Na^+ and Cl^- carry electrostatic charges depending on their valency. The electrostatic charge on chemical compounds is due to the presence of ionized groups such as $-NH_3^+$ and $-COO^-$. All amino acids carry at least one COOH group and one NH_2 group. Some amino acids have additional side chain groups. Whether an amino acid is charged or uncharged depends on the solution pH since it influences the extent of ionization. With proteins which are made up of large numbers of amino acids the situation is more complex. The electrostatic charge on a protein depends on the pK_a and pK_b of the individual constituent amino acids. Depending on the solution pH, a protein could have a net positive, neutral or negative charge, i.e. it is amphoteric in nature. At a pH value known as its isoelectric point, a protein has the same amount of positive and negative charges, i.e. is neutral in an overall sense. Above its isoelectric point a protein has a net negative charge while below this value it is has a net positive charge. Table 2.5 lists the isoelectric points of some proteins.

At physiological pH, nucleic acids are negatively charged. This is due to the presence of a large number of phosphate groups on these molecules. The electrostatic charge on molecules is the basis of separation in techniques such as electrophoresis, ion-exchange adsorption and chromatography, electrodialysis and precipitation.

Table 2.5 Isoelectric points of proteins

Protein	Isoelectric point
Ovalbumin	4.7
Conalbumin	7.1
Insulin	5.3
Urease	5.0
Lysozyme	11.0
Pepsin	1.0
Myoglobin	7.0
Collagen	6.7
Fibrinogen	4.8
Human serum albumin	4.9
Hemoglobin	7.1
Prolactin	5.7

2.8. Solubility

Solubility of a chemical substance in a standard solvent like water is one of its fundamental properties for characterization purposes. By rule of thumb, a polar compound will be more soluble in water than a non-polar compound. Also, a non-polar compound will be more soluble in an organic solvent than in water. The solubility of a substance can be influenced by the temperature, solution pH and the presence of additives. Solubility of a molecule is the basis of separation in techniques such as extraction, precipitation, crystallization and membrane separation. In precipitation based separation, the solubility of a substance is selectively decreased by manipulating one or more of the factors listed above. Generally speaking, the solubility of a substance in a liquid increases with increase in temperature. However, proteins denature at higher temperatures and precipitate in the form of a coagulated mass e.g. as in the poaching of eggs. From a separations point of view, precipitation will have to be reversible, i.e. we should be able to solubilize the precipitated substance by reversing the factors causing precipitation. The solubility of proteins is influenced by the presence of salts in solution. At very low ionic strengths, protein solubility is aided by salts, i.e. solubility increases with increase in salt concentration. This is referred to as the *salting in* effect. However, at higher ionic strengths, the solubility of proteins is found to decrease very significantly with increase in salt concentration. This is referred to as the *salting out* effect. The solution pH can also have a profound effect on the solubility of a protein. At its isoelectric point, a protein has its lowest solubility. On either sides of the isoelectric point, protein solubility is found to increase.

2.9. Partition coefficient

The partition coefficient is a measure of how a compound distributes itself between two liquid phases and is the basis of separation in processes such as liquid-liquid extraction and partition chromatography. This distribution of the compound is thermodynamically driven, the chemical potential of the compound in the two phases being equal at equilibrium. The ratio of the concentrations of the compound in the two phases at equilibrium is referred to as the partition coefficient. For organic compounds, the octanol/water partition coefficient ($K_{o/w}$) is used as a parameter to determine whether the compound is hydrophilic (water loving) or hydrophobic (water hating).

2.10. Light absorption

Solutions of different substances absorb light of different wavelengths. The wavelength at which a compound absorbs the maximum amount of light is referred to as its λ_{max}. Molecules which form colored solutions usually absorb visible light. Proteins in aqueous solutions absorb ultraviolet light, particularly at 280 nm wavelength while aqueous solutions of DNA absorb ultraviolet light preferably at 254 nm wavelength. The absorption of light is due to the presence of specific groups, or bonds within these molecules called chromophores. Light absorption is not a basis for separation but an important parameter by which to monitor compounds during a separation process e.g. as in liquid chromatography where the time at which different separated components leave the column are determined by measuring the light absorption of the column effluent. It is also an important tool by which the concentration and purity of substances can be determined e.g. as in spectrophotometry and HPLC.

The amount of light absorbed by a solution depends on its solute concentration and the path length of the light within the sample (see Fig. 2.4). The amount of light absorbed by a sample is quantified in terms of absorbance (A):

$$A = \log_{10}\left(\frac{I_i}{I_t}\right) \tag{2.8}$$

Where

I_i = intensity of the incident light

I_t = intensity of the transmitted light

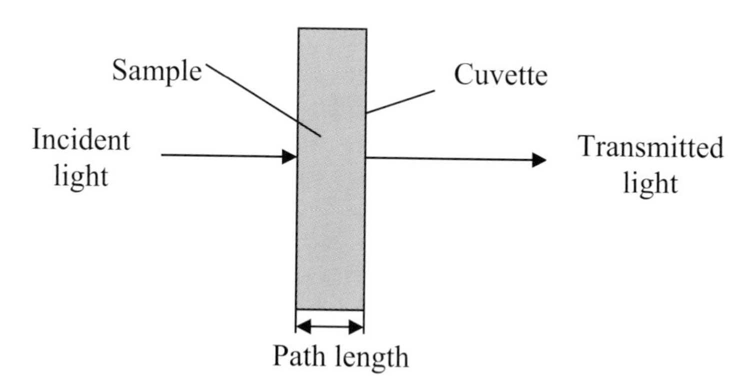

Fig. 2.4 Light absorption by sample in a cuvette

According to the Beer-Lambert law which holds good for dilute solutions:

$$A = aCl \tag{2.9}$$

Where

a = specific absorbance or absorptivity (AU m^2/kg)

C = concentration (kg/m^3)

l = path length (m)

2.11. Fluorescence

Certain compounds fluoresce, i.e. emit light after absorbing light of a higher frequency. Such substances are called fluorescent substances examples include proteins and nucleic acids. Specific compounds absorb and emit light of specific wavelengths. Fluorescence is not a basis for separation but an important parameter by which to monitor different substances during separation e.g. as in liquid chromatography, and immunoassays. It is also an important tool by which the concentration and purity of substances can be determined e.g. fluorimetry and HPLC. The emitted light in fluorimetry is measured at right angles to that of the incident light (Fig. 2.5) to avoid interference from the transmitted light. The intensity of emitted light can be correlated to the concentration.

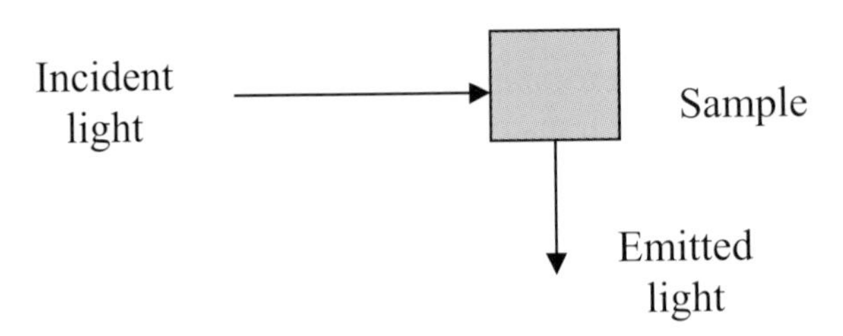

Fig. 2.5 Fluorescence

Exercise problems

2.1. Two spherical molecules A and B were found to have diffusivities of 4×10^{-10} m^2/s and 8×10^{-10} m^2/s respectively in a particular medium. Which molecule has the larger diameter and by what percent is this diameter greater than that of the other?

2.2. An ultrafiltration membrane separates two dilute myoglobin solution: 0.01 g/l and 0.05 g/l respectively, both being maintained at 25 degrees centigrade. Calculate the osmotic pressure across the membrane.

References

D.M. Freifelder, Physical Biochemistry: Applications to Biochemistry and Molecular Biology, 2nd edition, W.H. Freeman, New York (1982).

C. J. Geankoplis, Transport Processes and Separation Process Principles, 4th edition, Prentice Hall, Upper Saddle River (2003).

B.E. Poling, J.M. Prausnitz, J.P. O'Connell, The Properties of Gases and Liquids, 5th edition, McGraw Hill, New York (2000).

N.C. Price, R.A. Dwek, Principles and Problems in Physical Chemistry for Biochemists, 2nd edition, Oxford University Press, Oxford (1979).

O. Sten-Knudsen, Biological membranes: Theory of Transport, Potentials and Electric Impulses, Cambridge University Press, Cambridge (2002).

G.A. Truskey, F. Yuan, D.F. Katz, Transport Phenomena in Biological Systems, Pearson Prentice Hall, Upper Saddle River (2004).

Chapter 3

Mass transfer

3.1. Introduction

Any separation process involves three types of transport phenomena: heat transfer, fluid flow and mass transfer. Most bioseparation processes being isothermal or nearly isothermal in nature, the role of heat transfer is not as significant as in conventional chemical separations. The role of fluid flow is much more significant in comparison and will be dealt with in this text where relevant on a need to know basis. The role of mass transfer is perhaps most significant in bioseparations and hence an entire chapter is devoted to it.

Mass transfer as the term implies deals with transport of material. However, it is distinctly different from fluid flow which also deals with transport of material. A simple example of mass transfer is the movement of a scent from one end of a room to the other. Mass transfer basically deals with transport of species within a medium or across an interface, i.e. from one medium to another. The medium could be stationary or mobile. There are two types of mass transfer:

1. Purely diffusive mass transfer (or molecular diffusion)
2. Convective mass transfer

Molecular diffusion is governed by a random walk process and involves the transport of molecules from a region of high concentration to one where its concentration is lower. Steady state molecular diffusion of species *A* (e.g. sucrose) in a diffusion medium *B* (e.g. water) can be expressed by Fick's first law (see Fig. 3.1):

$$J_A = -D_{AB} \frac{dc_A}{dx} \tag{3.1}$$

Where

J_A = flux of *A* in *B* (kg-moles/m^2.s)

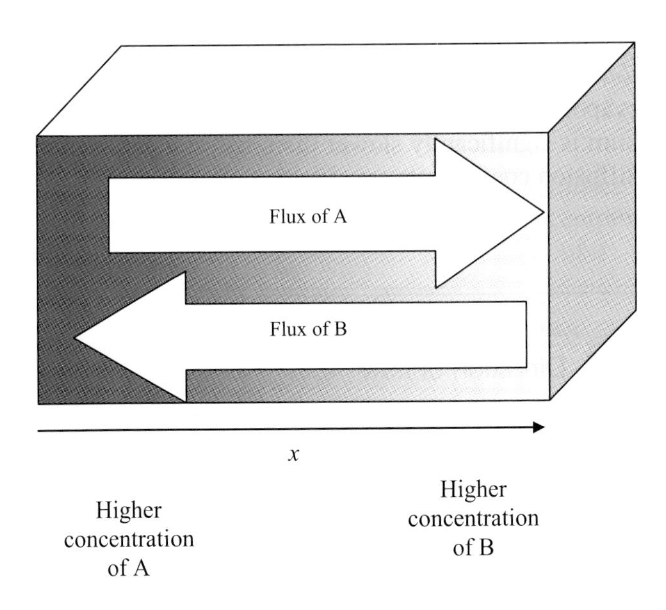

Fig. 3.3 Equimolar counter-diffusion

3.3. Measurement of diffusivity

The diffusivity of a solute in a liquid medium can be experimentally measured using various techniques. A commonly used technique is based on the diffusion cell (see Fig. 3.4) which consists of two well mixed chambers having the same volume which are separated by a porous membrane. Initially the two chambers are filled with the liquid medium. It is ensured that the pores of the membrane are also completely filled with the same liquid. At time $t = 0$, the liquid in one of the chambers (say chamber 1), is replaced with a solution of known concentration of the solute. The concentration of the solute in one or both chambers of the diffusion cell is/are then monitored, and based on the change in solute concentration with time its diffusivity can be determined using the following equation:

$$D_{AB} = \frac{V\delta\tau}{2\varepsilon at}\ln\left(\frac{c_1^0 - c_2^0}{c_1 - c_2}\right) \tag{3.7}$$

Where

V = volume of a chamber (m^3)

δ = thickness of the membrane (m)

τ = tortuosity of the membrane (-)
ε = porosity of the membrane (-)
a = area of the membrane (m^2)
t = time (s)
c_1 = solute concentration in chamber 1 (kg-mole/m^3)
c_2 = solute concentration in chamber 2 (kg-mole/m^3)

The superscript 0 represents initial value i.e. at $t = 0$.

There are several other techniques for measuring diffusivity, some of these being variants of the technique discussed above.

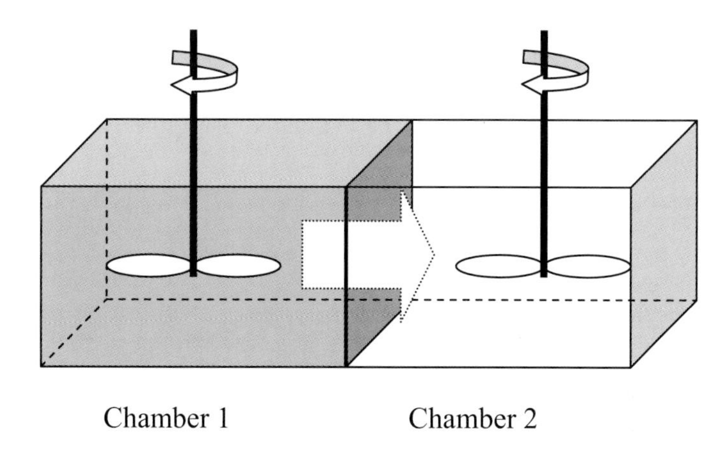

Fig. 3.4 Diffusion cell

3.4. Estimation of diffusivity

The diffusivity of a solute in a liquid medium at a particular temperature can also be estimated using different mathematical correlations. These correlations link diffusivity to solute and liquid properties such as molar volume, molecular weight and liquid viscosity. The three most widely used correlations are:

Stokes-Einstein correlation:

$$D_{AB} = \frac{9.96 \times 10^{-16} T}{\mu V_A^{1/3}} \tag{3.8}$$

Where

T = absolute temperature (K)
μ = viscosity if the liquid medium (kg/m s)

V_A = solute molar volume at its normal boiling point (m^3/kg-mole)

Wilke-Chang correlation:

$$D_{AB} = \frac{1.173 \times 10^{-16} (\phi M_B)^{1/2} T}{\mu V_A^{0.6}} \tag{3.9}$$

Where

ϕ = association parameter (-) and has a value of 2.6 for water

M_B = molecular weight of the liquid medium (kg/kg-mole)

Polson correlation:

$$D_{AB} = \frac{9.40 \times 10^{-15} T}{\mu M_A^{1/3}} \tag{3.10}$$

Where

M_A = molecular weight of the solute (kg/kg-mole)

Diffusivity of electrolytes can be estimated using the Nernst-Haskell correlation:

$$D_{AB} = \frac{8.928 \times 10^{-10} T(1/n_+ + 1/n_-)}{(1/\lambda_+ + 1/\lambda_-)} \tag{3.11}$$

Where

n_+ = valency of the cation

n_- = valency of the anion

λ_+ = ionic conductance of the cation

λ_- = ionic conductance of the anion

The correlation shown above gives diffusivity in cm^2/s. λ and n values for different cations and anions can be obtained from standard tables of physical properties.

The diffusion concepts discussed so far are based on simple systems, i.e. the solution of a single solute. Most systems handled in bioseparation processes are complex and hence the correlations discussed above have to be appropriately modified to account for specific system related effects.

Example 3.1

Estimate the diffusivity of the protein lysozyme in water at 25 degrees centigrade.

Solution

The diffusivity of a solute can be calculated from its molecular weight using Polson correlation i.e. equation (3.10). From Table 2.2 the molecular weight of lysozyme is 14,100 kg/kg-mole. The viscosity of water at 25 degrees centigrade is 0.001 kg/m s. Therefore:

$$D_{lysozyme} = \frac{9.40\times10^{-15}\times298}{0.001\times14100^{1/3}}\ \text{m}^2/\text{s} = 1.16\times10^{-10}\ \text{m}^2/\text{s}$$

3.5. Diffusion of solutes in dense solid

Solute molecules can diffuse through dense solid medium after dissolving in it. An example of this is the diffusion of ions through dense membranes. The molecules of the solid medium do not counter-diffuse on account of their limited mobility. However, Fick's law can still be used to describe the diffusion of solute molecules in a solid medium.

3.6. Diffusion of solutes in porous solid

Solute molecules can diffuse through the pores present in porous solids. In order for this to happen, the pores have to be filled with some liquid medium. Therefore no diffusion takes place through the solid material itself. All it does is hold the liquid medium in place. However, the solid material can have an influence on the diffusion within the liquid medium. It can for instance increase the effective diffusion path length of the solute if the pores are tortuous in nature. When the pores have dimension of the same order of magnitude as the solute, the pore wall can cause hindrance to diffusion. An example of un-hindered diffusion in a porous medium is the transport of sodium chloride through a microfiltration membrane (which has micron sized pores) while an example of hindered diffusion is the transport of albumin through an ultrafiltration membrane (which has nanometer sized pores). The steady state equation for un-hindered diffusion of a solute from point 1 and 2 within a slab of porous solid is given by:

$$J_A = \frac{\varepsilon D(c_{A1} - c_{A2})}{\tau(x_2 - x_1)} \qquad (3.12)$$

Where

D = diffusivity of the solute in the liquid within the pores (m^2/s)

ε = porosity of the medium (-)

τ = tortuosity of the medium (-)

The hindered diffusion of a solute through a porous solid from point 1 to 2 is given by:

$$J_A = \frac{D_{eff}\varepsilon(c_{A1} - c_{A2})}{\tau(x_2 - x_1)} \tag{3.13}$$

Where

D_{eff} = effective hindered diffusivity (m^2/s)

The effective hindered diffusivity of a solute in a pore can be obtained by:

$$D_{eff} = D\left(1 - \frac{d_s}{d_p}\right)^4 \tag{3.14}$$

Where

d_s = solute diameter (m)

d_p = pore diameter (m)

Example 3.2

A membrane having a porosity of 0.75, an average pore size of 1 micron, a surface area of 2 cm^2 and a thickness of 0.1 mm separates two water-filled, well-mixed chambers each having volume of 10 ml. The content of one of the chambers was replaced with 10 ml of 1 mg/ml human albumin solution at $t = 0$. Calculate the solute concentration in the other chamber after 50 minutes. Assume that the pores of the membrane are all aligned normal to the membrane surface and that there is no convective flow of solvent through the membrane.

Solution

In this problem we assume that the diffusion of albumin through the pores is not hindered. This can be confirmed by using equation (3.14). The diameter of albumin is 7.2×10^{-3} microns (ref. Table 2.1). Therefore:

$$D_{eff} = D$$

The diffusivity of albumin is 5.94×10^{-11} m^2/s (ref. Table 2.3). Using equation (3.7), we can write:

$$5.94\times10^{-11} = \frac{10\times10^{-6}\times0.1\times10^{-3}}{2\times0.75\times2\times10^{-4}\times3000}\ln\left(\frac{1}{C_1 - C_2}\right) \tag{3.a}$$

In this equation we have replaced the molar concentration with the mass concentration since the molecular weight cancels out between the numerator and denominator of the term within parenthesis. The total amount of solute (albumin) in the system remains constant. Therefore we can write:

$$C_1^0 V = C_1 V + C_2 V$$

Therefore:

$$1 \times 10 \times 10^{-6} = (C_1 + C_2) \times 10 \times 10^{-6} \qquad (3.b)$$

Solving equations (3.a) and (3.b) simultaneously, we get:

$C_1 = 0.973$ mg/ml

$C_2 = 0.027$ mg/ml

Example 3.3

Glucose is diffusing at 25 degrees centigrade in water within a porous medium having a porosity of 0.5, tortuosity of 1.8 and average pore diameter of 8.6× 10^{-3} microns. Determine the steady state flux of glucose between two points within the medium separated by a distance of 1 mm and having concentrations 1.5 g/l and 1.51 g/l respectively.

Solution

Table 2.2, gives the molecular weight of glucose as being 180 kg/kg-mole. The diffusivity of glucose at 25 degrees centigrade can be determined using Polson correlation i.e. equation (3.10):

$$D = \frac{9.40 \times 10^{-15} \times 298}{0.001 \times (180)^{1/3}} \text{ m}^2\text{/s} = 4.96 \times 10^{-10} \text{ m}^2\text{/s}$$

The size of glucose can be obtained from Table 2.1. The effective diffusivity of glucose in the porous structure can be obtained using equation (3.14):

$$D_{eff} = 4.96 \times 10^{-10} \left(1 - \frac{8.6 \times 10^{-4}}{8.6 \times 10^{-3}}\right)^4 \text{ m}^2\text{/s} = 3.25 \times 10^{-10} \text{ m}^2\text{/s}$$

The steady state flux of glucose can be obtained using equation (3.13):

$$J_A = \frac{3.25 \times 10^{-10} \times 0.5 \times 0.01}{1.8 \times 0.001 \times 180} \text{ kg-moles/m}^2\text{s} = 5.02 \times 10^{-12} \text{ kg-moles/m}^2\text{s}$$

3.7. Convective mass transfer

Convective mass transfer is observed in flowing fluids e.g. transport of a solute in a liquid flowing past a solid surface (see Fig. 3.5), or transport of a solute in a liquid flowing past another immiscible liquid (see Fig. 3.6). An example of the first type is the transfer of urea from blood towards the surface of a dialyser membrane in haemodialysis. An example of the second type is the transfer of penicillin G within filtered aqueous media flowing past an organic solvent in a liquid-liquid extractor. When a liquid flows past a solid surface a stagnant boundary liquid layer is formed close to the surface. Similarly when two liquids flow past one another, two boundary liquid layers are generated on either sides of the interface. Within these boundary layers, the transport of solute mainly takes place by molecular diffusion. If the flow of liquid is laminar, the transfer of solute in the directions indicated in Figs. 3.5 and 3.6 would be by molecular diffusion. However, if the flow were turbulent in nature, mass transfer would take place by a combination of molecular diffusion and eddy diffusion. This is referred to as convective mass transfer. The flux equation for convective mass transfer is:

$$N_A = -(D + E)\frac{dc_A}{dx} \tag{3.15}$$

Where

E = eddy diffusivity (m^2/s)

Equation (3.15) can be written as:

$$N_A = k_A \Delta c_A \tag{3.16}$$

Where

k_A = mass transfer coefficient (m/s)

Example 3.4

An aqueous solution of human immunoglobulin G (at 4 degrees centigrade) is being pumped through a tube having a diameter of 1 mm. The mass transfer coefficient for the protein in the radial direction was found to be 1×10^{-6} m/s. Comment on this value vis-à-vis the diffusivity of the protein. Estimate the eddy diffusivity of the flowing system.

Solution

From Table 2.2, the molecular weight of immunoglobulin G is 155,000 kg/kg-mole. Its diffusivity in water can be calculated using Polson correlation:

$$D_{AB} = \frac{9.40 \times 10^{-15} \times 277}{0.001 \times (155{,}000)^{1/3}} \text{ m}^2/\text{s} = 4.85 \times 10^{-11} \text{ m}^2/\text{s}$$

In this problem the diffusion length is the radius of the tube which is 0.5 mm. If solute transport is due to diffusion alone, the mass transfer coefficient is obtained by dividing the diffusivity by the diffusion length. Therefore:

$$k_D = \frac{4.85\times10^{-11}}{0.5\times10^{-3}}\,\text{m/s} = 9.69 \times 10^{-8}\,\text{m/s}$$

This is significantly lower than the observed mass transfer coefficient. Therefore there is some eddy diffusivity involved in solute transport. Eddy diffusivity can be obtained from equations (3.15) and (3.16):

$$E = (1\times10^{-6}\times0.5\times10^{-3}) - 4.85\times10^{-11}\ \text{m}^2/\text{s} = 4.52 \times 10^{-10}\ \text{m}^2/\text{s}$$

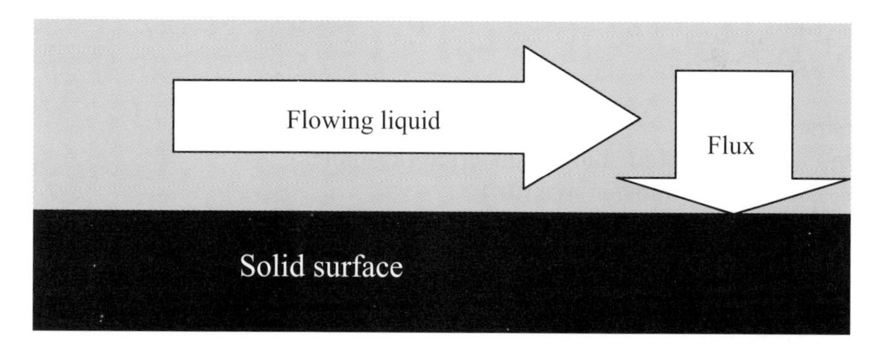

Fig. 3.5 Convective mass transfer in solid-liquid system

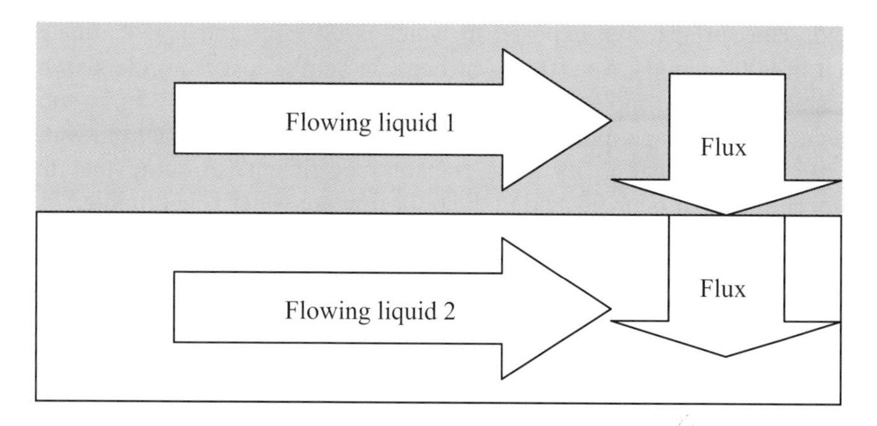

Fig. 3.6 Convective mass transfer in liquid-liquid system

3.8. Experimental determination of mass transfer coefficient

Mass transfer coefficient of a solute in a flowing liquid can be determined by carrying out steady-state experiments based on equation (3.16). The general approach in these experiments is to measure the amount of solute transferred in a given period of time across a given surface area for a particular concentration difference across the transfer zone. Solute mass transfer coefficient in a liquid flowing within a tube can be determined by coating the solute over the inner wall of the tube followed by measurement of the amount of solute removed. The average flux across the mass transfer zone can be calculated from:

$$N_A = \frac{m}{a\,\Delta t} \tag{3.17}$$

Where

m = amount of solute transferred (kg-moles)

a = mass transfer area (m^2)

Δt = transfer time (s)

From equations (3.16) and (3.17):

$$k_A = \frac{m}{a\;\Delta t\;\Delta c_A} \tag{3.18}$$

Example 3.5

Water is flowing past one of the rectangular sides of a slab of benzoic acid. The surface area exposed to water is 0.01 m^2 and it is estimated that in 600 seconds, 5×10^{-4} kg of benzoic acid is lost from the slab by dissolution. If the molecular weight of benzoic acid is 121.1 kg/kg-mol, calculate its average molar flux. If the solubility of benzoic acid in water is 0.2 kg/m^3, what is its mass transfer coefficient? Assume that the concentration of benzoic acid in the bulk flowing water is negligible.

Solution

The amount of solute transferred in 600 s is:

$$m = \frac{5\times10^{-4}}{121.1}\text{ kg-moles} = 4.13 \times 10^{-6}\text{ kg-moles}$$

The flux can be calculated using equation (3.17):

$$N_A = \frac{4.13\times10^{-6}}{0.01\times600}\text{ kg-moles/m}^2\text{ s} = 6.9 \times 10^{-7}\text{ kg-moles/m}^2\text{ s}$$

Assuming that the benzoic acid concentration on the surface of the slab is the same as its solubility, the concentration difference across the transfer zone is:

$$\Delta c = \frac{0.2}{121.1} \text{ kg-moles/m}^3 = 1.65 \times 10^{-3} \text{ kg-moles/m}^3$$

The mass transfer coefficient can be calculated using equation (3.18):

$$k_A = \frac{4.13 \times 10^{-6}}{0.01 \times 600 \times 1.65 \times 10^{-3}} \text{ m/s} = 4.17 \times 10^{-4} \text{ m/s}$$

3.9. Estimation of mass transfer coefficient

The mass transfer coefficient can be estimated using numerical correlations which are based on heat-mass transfer analogy. These correlations typically have three or more dimensionless groups. The three dimensionless groups that are always present in these correlations are the Reynolds number (N_{Re}), the Sherwood number (N_{Sh}) and the Schmidt number (N_{Sc}):

$$N_{Re} = \frac{du\rho}{\mu} \tag{3.19}$$

Where

d = hydraulic diameter of the flow passage (m), e.g. tube diameter

u = velocity of flowing liquid (m/s)

ρ = density (kg/m^3)

μ = solution viscosity (kg/m s)

$$N_{Sh} = \frac{k_A d}{D} \tag{3.20}$$

Where

D = solute diffusivity (m^2/s)

$$N_{Sc} = \frac{\mu}{\rho D} \tag{3.21}$$

The general form of such dimensionless correlations is:

$$N_{Sh} = a\, N_{Re}^{b} N_{Sc}^{c} \tag{3.22}$$

Where a, b and c are constants

Some dimensionless correlations of this type will be discussed in the chapter on *membrane based bioseparation*.

3.10. Inter-phase mass transfer

So far we have discussed mass transfer within a medium. The transport of a solute from one medium to another across an interface is called inter-phase or interfacial mass transfer. Such type of material transport is quite common in separation processes such as liquid-liquid extraction, leaching, chromatography and membrane separation. As an example of inter-phase mass transfer, the transport of a solute from a liquid to another immiscible liquid will be discussed here. For the sake of simplicity the two liquids are assumed to be stagnant thus eliminating the need for considering boundary layers.

Fig. 3.7 shows the steady-state solute concentration profile close to the interface between the two liquids. The solute flux across the interface which takes place from liquid 1 to liquid 2 is given by:

$$J = D_1 \frac{(c_1 - c_{i1})}{x_1} = D_2 \frac{(c_{i2} - c_2)}{x_2} \tag{3.23}$$

Where

c_1 = solute concentration at x_1 distance from interface in liquid 1
c_2 = solute concentration at x_2 distance from interface in liquid 2
c_{i1} = interfacial solute concentration in liquid 1
c_{i2} = interfacial solute concentration in liquid 2
D_1 = diffusivity in liquid 1
D_2 = diffusivity in liquid 2

The interfacial solute concentrations in the two liquids are linked by the partition coefficient:

$$c_{i2} = K\, c_{i1} \tag{3.24}$$

Where

K = partition coefficient (-)

3.11. Unsteady state mass transfer

The discussion so far has been based on steady-state mass transfer, i.e. where the concentrations at various locations within the transfer zone do not change with time. However, in many separations, particularly in rate processes, these concentrations do change with time and hence it is important to understand what happens in an unsteady state mass transfer process.

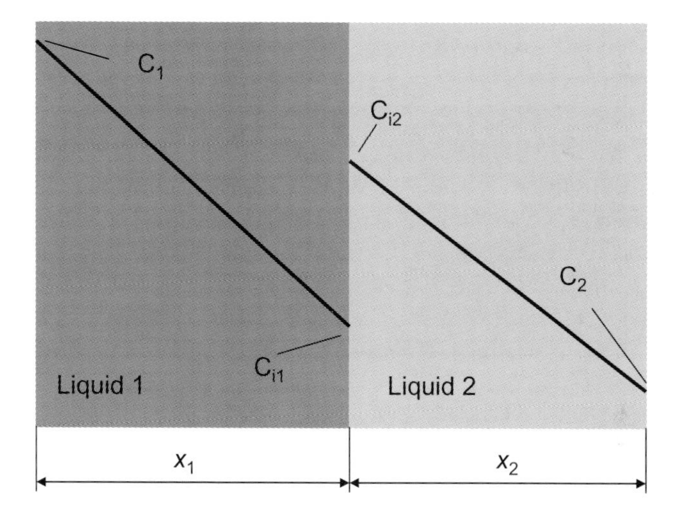

Fig. 3.7 Solute concentration profile across interface

The general equation for unsteady state molecular diffusion in 3-dimension for the Cartesian co-ordinate system is of the form shown below:

$$D_{AB}\left(\frac{\partial^2 c_A}{\partial x^2}+\frac{\partial^2 c_A}{\partial y^2}+\frac{\partial^2 c_A}{\partial z^2}\right)=\frac{\partial c_A}{\partial t} \qquad (3.25)$$

Depending on the geometry of the system under consideration, an appropriate partial differential equation which could be a modified form of equation (3.25) needs to be set up. This is then solved taking into consideration appropriate initial and boundary conditions. As a case study, the unsteady state diffusion of a solute in a semi-infinite gel is discussed here. If a solution having a solute concentration c_{A1} is suddenly brought into contact with a thick slab of gel within which the initial concentration of the solute is c_{A0}, the concentrations within the slab at various locations will change with time as shown in Fig. 3.8. In order to simplify the problem we assume that the concentration of the solute in the solution adjacent to the slab remains constant. We also assume that the solute concentration is identical on either sides of the solution-slab interface, i.e. its partition coefficient is unity. This is an example of mass transfer in 1-dimension, i.e. along the x-axis. Equation (3.25) reduces to:

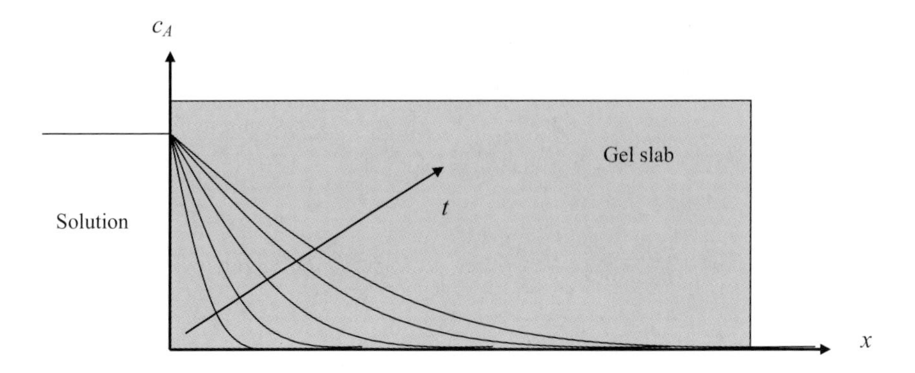

Fig. 3.8 Unsteady state diffusion in semi-infinite slab

$$D_{AB}\frac{\partial^2 c_A}{\partial x^2}=\frac{\partial c_A}{\partial t} \tag{3.26}$$

Initial condition is:
At $t = 0$, for all x, $c_A = c_{A0}$

Boundary conditions are:
At $x = 0$, for $t > 0$, $c_A= c_{A1}$ i.e. $c_A(0,t) = c_{A1}$
At $x =\infty$, for all t, $c_A = c_{A0}$, i.e. $c_A(\infty,t) = c_{A0}$

Solving equation (3.26) we get:

$$\frac{c_{A1}-c_A}{c_{A1}-c_{A0}}=erf\left(\frac{x}{2\sqrt{D_{AB}t}}\right) \tag{3.27}$$

Using this equation the solute concentration at any point x at any time t can be calculated. The flux of solute across the interface at any time t is given by:

$$J_A(0,t)=\sqrt{\frac{D_{AB}}{\pi t}}\left(c_{A1}-c_{A0}\right) \tag{3.28}$$

Example 3.6
A thick gel slab of agar is suddenly exposed on one side to a well-mixed aqueous solution of an antibiotic, the purpose of this being to imbibe the antibiotic within the gel. The concentration of the antibiotic solution is 0.001 kg-moles/m^3. If we assume that the concentration in the well-mixed solution does not change appreciably in the time during which this experiment is carried out determine the flux of the antibiotic across the

gel-water interface after 5 minutes, given that its diffusivity in the gel is 8×10^{-11} m^2/s.

Solution

The flux of the antibiotic across the interface can be calculated using equation (3.28):

$$J_A(0,600) = \sqrt{\frac{8 \times 10^{-11}}{3.142 \times 300}}(0.001 - 0) \text{ kg-moles/m}^2 \text{ s}$$

$= 2.91 \times 10^{-11}$ kg-moles/m^2 s

3.12. Equilibrium and rate processes

Transfer of species from one zone of a continuum to another or indeed across an interface continues to take place until some form of equilibrium is established. This equilibrium could be defined in terms of chemical potential, concentration or other appropriate parameters. Some separation processes are carried out until equilibrium has been achieved e.g. liquid-liquid extraction of penicillin G from fermentation media to methyl isobutyl ketone. Such processes are referred to as equilibrium processes (Fig. 3.9).

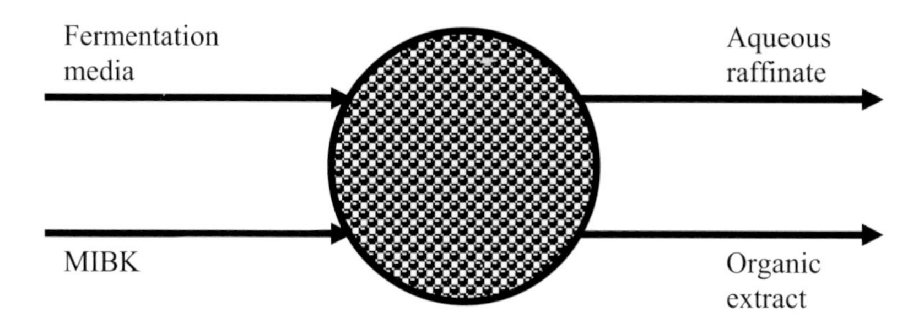

Fig. 3.9 Equilibrium process: Extraction of penicillin G

However in many separations, the nature of the process is such that no identifiable equilibrium is reached. An example of this is haemodialysis where maintenance of a concentration gradient throughout the process is essential. Such processes are referred to as non-equilibrium or rate processes (Fig. 3.10).

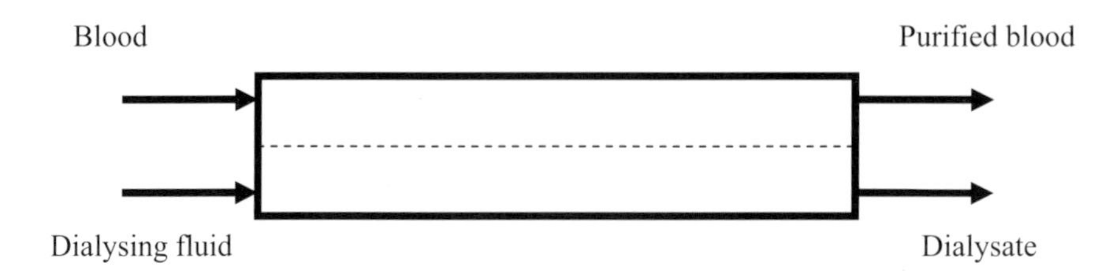

Fig. 3.10 Rate process: Haemodialysis

Exercise problems

3.1. An aqueous solution of lysozyme (MW 14,100 kg/kg-mole) at a temperature of 25 degrees centigrade is being pumped at a flow rate of 5000 ml/min through a short tube having a diameter of 10 mm. The concentration of lysozyme in the feed solution is 1 g/l while its concentration adjacent to the tube wall is practically zero on account of the rapid adsorption of the protein on the wall of the tube. The steady state flux of lysozyme in the radial direction within the tube was found to be 1.3×10^{-11} kg-mole/m^2.s. If molecular diffusion alone would have taken place, what would have been the steady state flux in the radial direction?

3.2. A membrane having a porosity of 0.3, average pore size of 0.01 micron, average pore tortuosity of 1.1, surface area of 2 cm^2 and thickness of 0.1 mm separates two water-filled, well-mixed chambers one having a volume of 10 ml and the other having a volume of 20 ml. The content of the 10 ml chamber was replaced with 10 ml of 1 mg/ml human albumin solution at $t = 0$. Calculate the albumin concentration in the other chamber after 300 minutes. Assume that there is no convective flow of solvent through the membrane. In this problem you cannot assume that the concentration in the 10 ml chamber does not change significantly with time. This experiment was carried out at 20 degrees centigrade.

3.3. A thick slab of agar containing 0.05 g/l of lysine is prepared in a Petri dish. The portion on top of the gel was suddenly flooded with a 1 g/l lysine solution. If we assume that the amino acid concentration in the solution does not change significantly with time determine the flux of lysine across the gel-water interface after

600 seconds. Predict the concentration of lysine at a distance of 5 mm from the interface at that time, given that its diffusivity in the gel is 6.5×10^{-9} m^2/s.

3.4. Human immunoglobulin G (HIgG) is diffusing through a porous medium having a porosity of 0.5, tortuosity of 1.8 and average pore diameter of 0.45 microns. Protein samples were collected from two points within the medium separated by a distance of 20 mm and the ultraviolet light absorbance of these samples as measured with a spectrophotometer having a sample path length of 0.2 cm were found to be 0.010 and 0.012 respectively. If the specific absorbance of HIgG is known to be 15590 m^3/kg-moles.cm, calculate the steady state flux of this protein between the two points.

References

R.B. Bird, W.E. Stewart, E.N. Lightfoot, Transport Phenomena, 2nd Edition, John Wiley and Sons, New York (2002).

J. Crank, The Mathematics of Diffusion, 2nd edition, Oxford University Press, Oxford (1980).

E.L. Cussler, Diffusion: Mass Transfer in Fluid Systems, Cambridge University Press, Cambridge (1997).

C. J. Geankoplis, Transport Processes and Separation Process Principles, 4th edition, Prentice Hall, Upper Saddle River (2003).

W.L. McCabe, J.C. Smith, P. Harriott, Unit Operations of Chemical Engineering, 7th edition, McGraw Hill, New York (2005).

R.F. Probstein, Physicochemical Hydrodynamics: An Introduction, 2nd edition, John Wiley and Sons (2003).

O. Sten-Knudsen, Biological membranes: Theory of Transport, Potentials and Electric Impulses, Cambridge University Press, Cambridge (2002).

G.A. Truskey, Fan Yuan, D.F. Katz, Transport Phenomena in Biological Systems, Pearson Prentice Hall, Upper Saddle River (2004).

Chapter 4

Cell disruption

4.1. Introduction

Biological products synthesized by fermentation or cell culture are either intracellular or extracellular. Intracellular products either occur in a soluble form in the cytoplasm or are produced as inclusion bodies (fine particles deposited within the cells). Examples of intracellular products include recombinant insulin and recombinant growth factors. A large number of recombinant products form inclusion bodies in order to accumulate in larger quantities within the cells. In order to obtain intracellular products the cells first have to be disrupted to release these into a liquid medium before further separation can be carried out. Certain biological products have to be extracted from tissues, an example being porcine insulin which is obtained from pig pancreas. In order to obtain such a tissue-derived substance, the source tissue first needs to be homogenized or ground into a cellular suspension and the cells are then subjected to cell disruption to release the product into a solution. In the manufacturing process for intracellular products, the cells are usually first separated from the culture liquid medium. This is done in order to reduce the amount of impurity: particularly secreted extracellular substances and unutilized media components. In many cases the cell suspensions are thickened or concentrated by microfiltration or centrifugation in order to reduce the process volume.

4.2. Cells

Different types of cell need to be disrupted in the bio-industry:

1. Gram positive bacterial cells
2. Gram negative bacterial cells
3. Yeast cell

4. Mould cells
5. Cultured mammalian cells
6. Cultured plant cells
7. Ground tissue

Fig. 4.1 shows the barriers present in a gram positive bacteria. The main barrier is the cell wall which is composed of peptidoglycan, teichoic acid and polysaccharides and is about 0.02 to 0.04 microns thick. The plasma or cell membrane which is made up of phospholipids and proteins is relatively fragile. In certain cases polysaccharide capsules may be present outside the cell wall. The cell wall of gram positive bacteria is particularly susceptible to lysis by the antibacterial enzyme lysozyme.

Fig. 4.2 shows the barriers present in a gram negative bacteria. Unlike gram positive bacteria these do not have distinct cell walls but instead have multi-layered envelops. The peptidoglycan layer is significantly thinner than in gram positive bacteria. An external layer composed of lipopolysaccharides and proteins is usually present. Another difference with gram positive bacteria is the presence of the periplasm layers which are two liquid filled gaps, one between the plasma membrane and the peptidoglycan layer and the other between the peptidoglycan layer and the external lipopolysaccharides. Periplasmic layers also exits in gram positive bacteria but these are significantly thinner than those in gram negative bacteria. The periplasm is important in bioprocessing since a large number of proteins, particularly recombinant proteins are secreted into it. An elegant way to recover the periplasmic proteins is by the use of osmotic shock. This technique is discussed at the end of the chapter.

Yeasts which are unicellular have thick cell walls, typically 0.1 to 0.2 microns in thickness. These are mainly composed of polysaccharides such as glucans, mannans and chitins. The plasma membrane in a yeast cell is composed of phospholipids and lipoproteins. Mould cells are largely similar to yeast cells in terms of cell wall and plasma membrane composition but are multicellular and filamentous. Mammalian cells do not possess the cell wall and are hence quite fragile i.e. easy to disrupt. Plant cells on the other hand have very thick cell walls mainly composed of cellulose and other polysaccharides.

Cell wall wherever present is the main barrier which needs to be disrupted to recover intracellular products. A range of mechanical methods can be used to disrupt the cell wall. Chemical methods when used for cell disruption are based on specific targeting of key cell wall components. For instance, lysozyme is used to disrupt the cell wall of gram positive bacteria since it degrades peptidoglycan which is a key cell wall constituent. In gram negative bacteria, the peptidoglycan layer is less susceptible to lysis by lysozyme since it is shielded by a layer composed of lipopolysaccharide and protein.

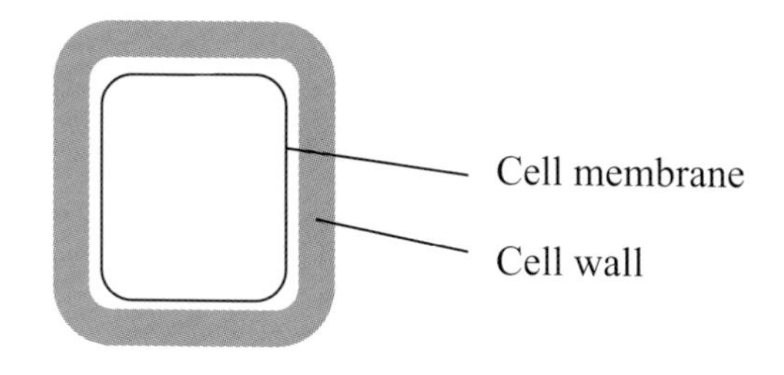

Fig. 4.1 Gram positive bacteria

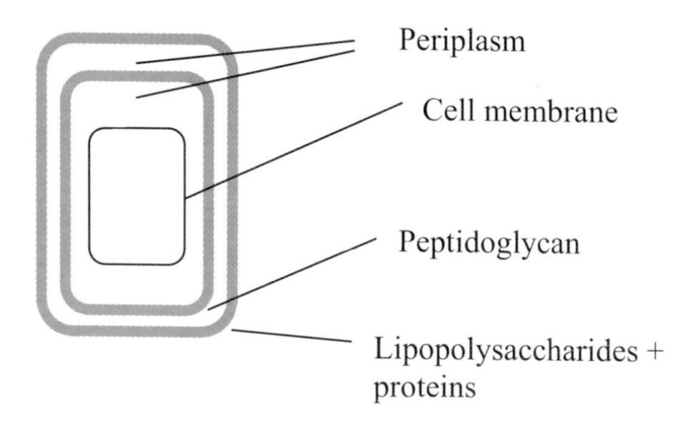

Fig. 4.2 Gram negative bacteria

Cell membranes or plasma membranes are composed of phospholipids arranged in the form of a bi-layer with the hydrophilic groups of the phospholipids molecules facing outside (see Fig. 4.3). The hydrophobic residues remain inside the cell membrane where they are shielded from the aqueous environment present both within and outside the cell. The plasma membrane can be easily destabilized by detergents, acid, alkali and organic solvents. The plasma membrane is also quite fragile when compared to the cell wall and can easily be disrupted using osmotic shock i.e. by suddenly changing the osmotic pressure across the membrane. This can be achieved simply by transferring the cell from isotonic medium to distilled water.

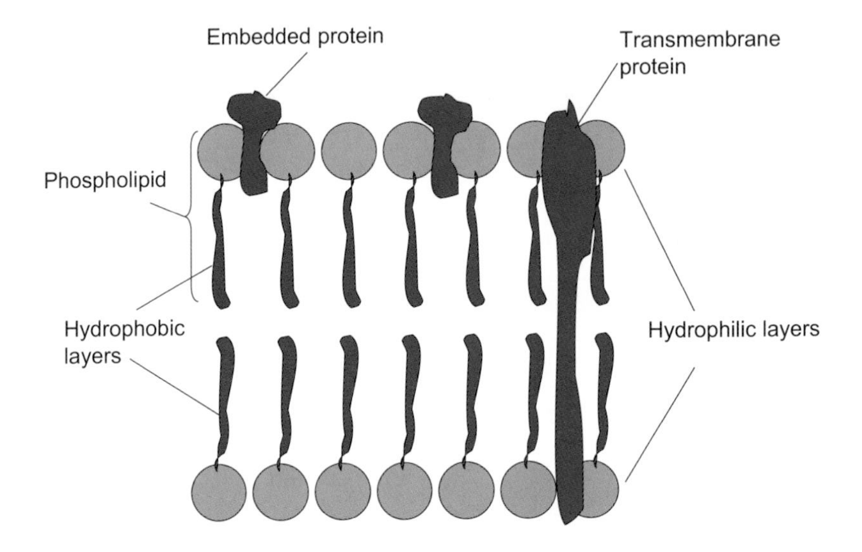

Fig. 4.3 Cell membrane

Cell disruption methods can be classified into two categories: physical methods and chemical methods.

Physical methods

1. Disruption in bead mill
2. Disruption using a rotor-stator mill
3. Disruption using French press
4. Disruption using ultrasonic vibrations

Chemical and physicochemical methods

1. Disruption using detergents
2. Disruption using enzymes e.g. lysozyme
3. Disruption using solvents
4. Disruption using osmotic shock

The physical methods are targeted more towards cell wall disruption while the chemical and physicochemical methods are mainly used for destabilizing the cell membrane.

4.3. Cell disruption using bead mill

Fig. 4.4 illustrates the principle of cell disruption using a bead mill. This equipment consists of a tubular vessel made of metal or thick glass within which the cell suspension is placed along with small metal or glass beads. The tubular vessel is then rotated about its axis and as a result of this the beads start rolling away from the direction of the vessel rotation. At higher rotation speeds, some the beads move up along with the curved wall of the vessel and then cascade back on the mass of beads and cells below. The cell disruption takes place due to the grinding action of the rolling beads as well as the impact resulting from the cascading beads.

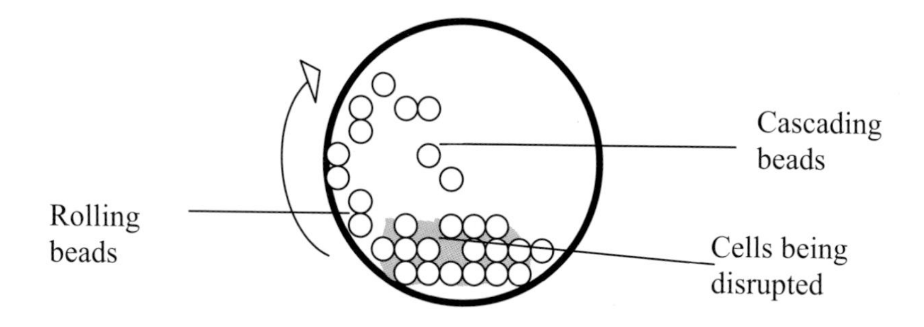

Fig. 4.4 Cell disruption using bead mill

Bead milling can generate enormous amounts of heat. While processing thermolabile material, the milling can be carried out at low temperatures, i.e. by adding a little liquid nitrogen into the vessel. This is referred to as cryogenic bead milling. An alternative approach is to use glycol cooled equipment. A bead mill can be operated in a batch mode or in a continuous mode and is commonly used for disrupting yeast cells and for grinding animal tissue. Using a small scale unit operated in a continuous mode, a few kilograms of yeast cells can be disrupted per hour. Larger unit can handle hundreds of kilograms of cells per hour. Fig. 4.5 shows an industrial scale bead mill.

Cell disruption involves particle size reduction and has certain similarities with grinding. According to the Kick's law of grinding, the amount of energy required to reduce the size of material is proportional to the size reduction ratio:

$$\frac{dE}{dr} = \frac{K_K f_c}{r} \tag{4.1}$$

Where

r = radius of the particle (m)

E = energy (Joules)

K_K = Kick's coefficient

f_c = crushing strength of the material

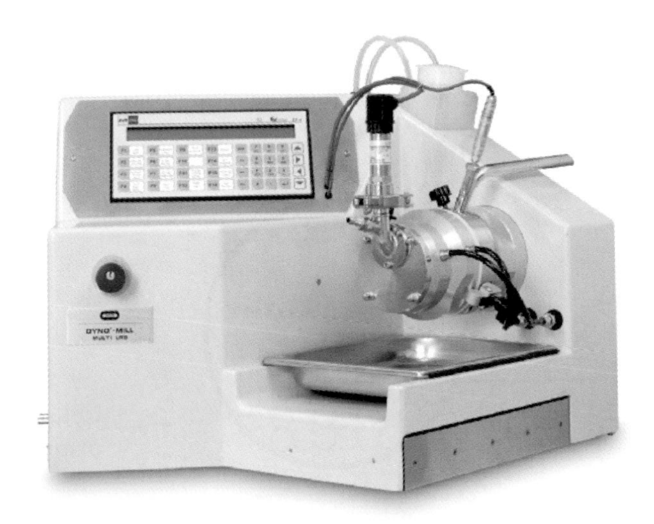

Fig. 4.5 Bead mill (Photo courtesy of Glen Mills Inc. and W. A. Bachofen AG)

The Kick's coefficient depends on the equipment and operating conditions. Integrating equation (4.1) for a given size reduction i.e. from r_1 to r_2:

$$E = K_K f_c \ln\left(\frac{r_1}{r_2}\right) \tag{4.2}$$

According to the Rittinger's law of grinding, the amount of energy needed for size reduction is proportional to the change in surface area:

$$\frac{dE}{dr} = \frac{K_R f_c}{r^2} \tag{4.3}$$

Where

K_R = Rittinger's coefficient

The Rittinger's coefficient depends on the process and equipment. Integrating equation (4.3) for a given size reduction:

$$E = K_R f_c \left(\frac{1}{r_2} - \frac{1}{r_1} \right) \tag{4.4}$$

Example 4.1

A bead mill was used to grind *Penicillium* filaments and the energy required for different size reductions for the same mass of material was determined (see Table below):

Average initial radius (microns)	Average final radius (microns)	Energy required (J)
6	5.5	1.8
5	4.5	2.7
4	3.5	4.3
3	2.5	8.0
2	1.5	20.0

Calculate the amount of energy required to reduce the average filament radius from 5 microns to 1 micron for the same mass of *Penicillium* as used in the above study in the same bead mill.

Solution

If the grinding process described above follows Kick's law a plot of energy versus ln (r_1/r_2) should give us a straight line. Similarly if the grinding process follows Rittinger's law, a plot of energy versus $((1/r_2)$-$(1/r_1))$ should give us a straight line. The second plot gives a significantly better fit, clearly indicating that the grinding process follows Rittinger's law. From this plot:

$$K_R f_c = 120.03 \times 10^{-6} \text{ J m}$$

Therefore the energy required to reduce the radius from 5 microns to 1 micron is:

$$E = 120.03 \times 10^{-6} \left(\frac{1}{1 \times 10^{-6}} - \frac{1}{5 \times 10^{-6}} \right) \text{J} = 96.02 \text{ J}$$

Kick's law and Rittinger's law are better suited for tissue grinding. Cell disruption primarily involves breaking the barriers around the cells followed by release of soluble and particulate sub-cellular components into the external liquid medium. This is a random process and hence incredibly hard to model. Empirical models are therefore more often used for cell disruption:

$$\frac{C}{C_{\max}} = 1 - \exp\left(-\frac{t}{\theta}\right) \qquad (4.5)$$

Where

C = concentration of released product (kg/m^3)

C_{max} = maximum concentration of released material (kg/m^3)

t = time (s)

θ = time constant (s)

The time constant θ depends on the processing conditions, equipment and the properties of the cells being disrupted.

4.4. Cell disruption using rotor-stator mill

Fig. 4.6 shows the principle of cell disruption using a rotor-stator mill. This device consists of a stationary block with a tapered cavity called the stator and a truncated cone shaped rotating object called the rotor. Typical rotation speeds are in the 10,000 to 50,000 rpm range. The cell suspension is fed into the tiny gap between the rotating rotor and the fixed stator. The feed is drawn in due to the rotation and expelled through the outlet due to centrifugal action. The high rate of shear generated in the space between the rotor and the stator as well as the turbulence thus generated are responsible for cell disruption. Fig. 4.7 shows a laboratory scale rotor-stator cell disruption mill. These mills are more commonly used for disruption of plant and animal tissues based material and are operated in the multi-pass mode, i.e. the disrupted material is sent back into the device for more complete disruption. The cell disruption caused within the rotor-stator mill can be described using the equations discussed for a bead mill. In a multi-pass operation:

$$\frac{C}{C_{\max}} = \left(1 - \exp\left(-\frac{t}{\theta}\right)\right)^{N} \qquad (4.6)$$

Where

N is the number of passes (-)

4.5. Cell disruption using French press

Fig. 4.8 shows the working principle of a French press which is a device commonly used for small-scale recovery of intracellular proteins and DNA from bacterial and plant cells. The device consists of a cylinder fitted with a plunger which is connected to a hydraulic press. The cell suspension is placed within the cylinder and pressurized using the plunger. The cylinder is provided with an orifice through which the suspension emerges at very high velocity in the form of a fine jet. The cell disruption takes place primarily due to the high shear rates influence by the cells within the orifice. A French press is frequently provided with an impact plate, where the jet impinges causing further cell disruption. Typical volumes handled by such devices range from a few millilitres to a few hundred millilitres. Typical operating pressure ranges from 10,000 to 50,000 psig.

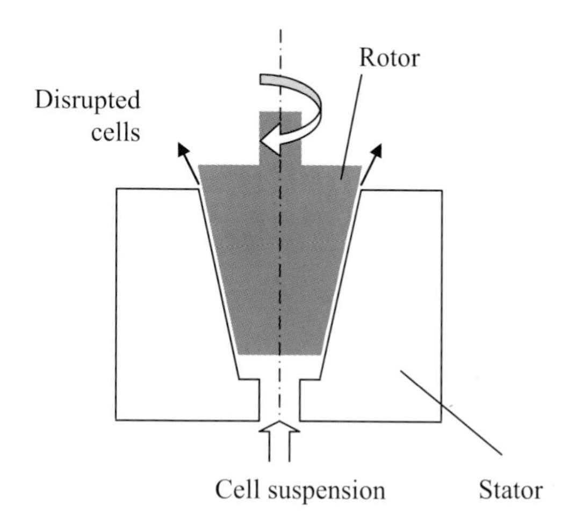

Fig. 4.6 Cell disruption using rotor-stator mill

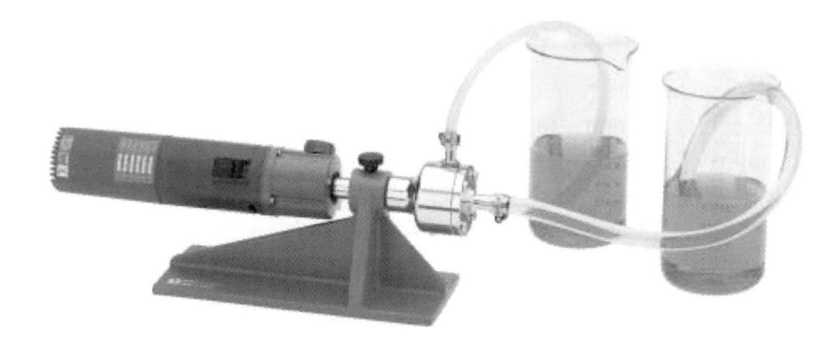

Fig. 4.7 Flow-through rotor-stator mill (Photo courtesy of IKA® -WERKE GmbH & Co. KG)

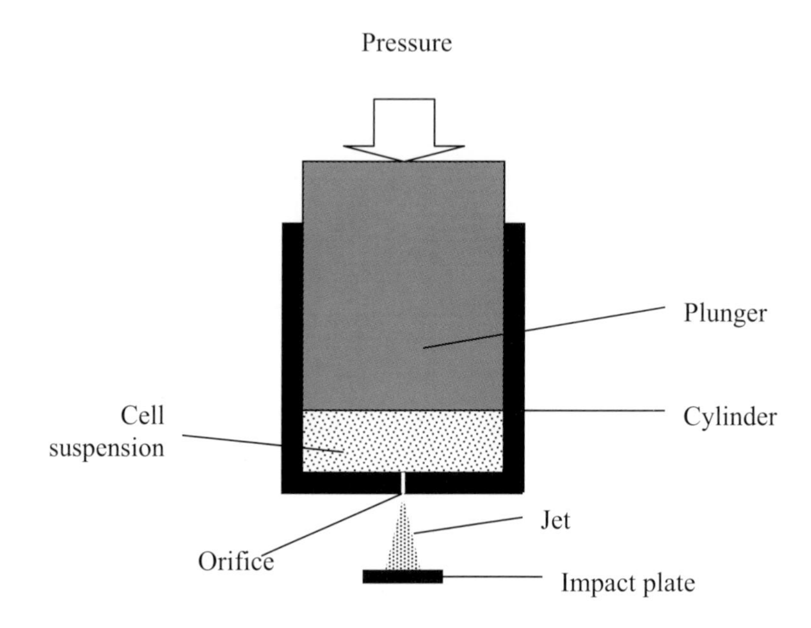

Fig. 4.8 Cell disruption using a French press

4.6. Cell disruption using ultrasonic vibrations

Ultrasonic vibrations (i.e. having frequency greater than 18 kHz) can be used to disrupt cells. The cells are subjected to ultrasonic vibrations by introducing an ultrasonic vibration emitting tip into the cell suspension (Fig. 4.9). Ultrasound emitting tips of various sizes are available and these are selected based on the volume of sample being processed. The ultrasonic vibration could be emitted continuously or in the form of short pulses. A frequency of 25 kHz is commonly used for cell disruption. The duration of ultrasound needed depends on the cell type, the sample size and the cell concentration. These high frequency vibrations cause cavitations, i.e. the formation of tiny bubbles within the liquid medium (see Fig. 4.10). When these bubbles reach resonance size, they collapse releasing mechanical energy in the form of shock waves equivalent to several thousand atmospheres of pressure. The shock waves disrupts cells present in suspension. For bacterial cells such as *E. coli*, 30 to 60 seconds may be sufficient for small samples. For yeast cells, this duration could be anything from 2 to 10 minutes. Fig. 4.11 shows a laboratory scale ultrasonic cell disrupter.

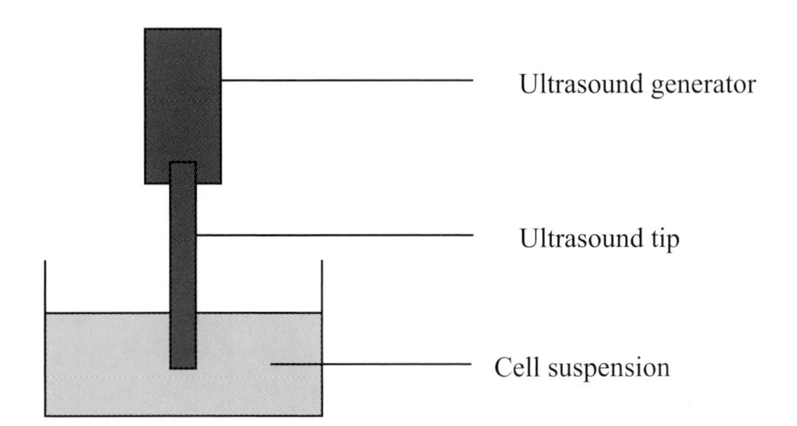

Fig. 4.9 Ultrasonic cell disruption

Fig. 4.10 Cavitation during ultrasonic cell disruption (Photo courtesy of Glen Mills Inc.)

Ultrasonic vibration is frequently used in conjunction with chemical cell disruption methods. In such cases the barriers around the cells are first weakened by exposing them to small amounts of enzymes or detergents. Using this approach, the amount of energy needed for cell disruption is significantly reduced.

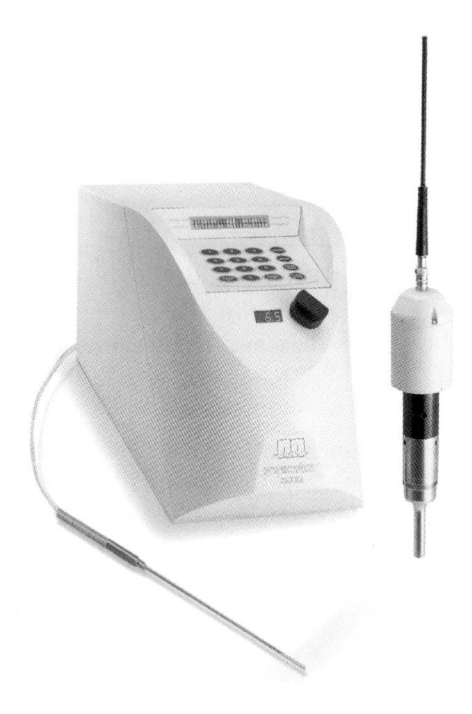

Fig. 4.11 Laboratory scale ultrasonic disrupter (Photo courtesy of Glen Mills Inc.)

Example 4.2

A batch of yeast cells was disrupted using ultrasonic vibrations to release an intracellular product. The concentration of released product in the solution was measured during the process (see table below):

Time (s)	Concentration (mg/ml)
60	3.49
120	4.56

If the ultrasonic cell disruption were carried out for 240 seconds, predict the product concentration.

Solution

Based on equation (4.5) we can write:

$$3.49 = C_{max}\left(1 - \exp\left(-\frac{60}{\theta}\right)\right)$$

$$4.56 = C_{max}\left(1 - \exp\left(-\frac{120}{\theta}\right)\right)$$

Solving these equations simultaneously we get:

C_{max} = 5 mg/ml
θ = 50 s

Therefore when t = 240 s:

$$C = 5\left(1 - \exp\left(-\frac{240}{50}\right)\right) \text{ mg/ml} = 4.96 \text{ mg/ml}$$

4.7. Cell disruption using detergents

Detergents disrupt the structure of cell membranes by solubilizing their phospholipids. These chemicals are mainly used to rupture mammalian cells. For disrupting bacterial cells, detergents have to be used in conjunction with lysozyme. With fungal cells (i.e. yeast and mould) the cell walls have to be similarly weakened before detergents can act. Detergents are classified into three categories: cationic, anionic and non-ionic. Non-ionic detergents are preferred in bioprocessing since they cause the least amount of damage to sensitive biological molecules such as proteins and DNA. Commonly used non-ionic detergents include the Triton-X series and the Tween series. However, it must be noted that a large number of proteins denature or precipitate in presence of detergents. Also, the detergent needs to be subsequently removed from the product and this usually involves an additional purification/polishing step in the process. Hence the use of detergents is avoided where possible.

4.8. Cell disruption using enzymes

Lysozyme (an egg based enzyme) lyses bacterial cell walls, mainly those of the gram positive type. Lysozyme on its own cannot disrupt bacterial cells since it does not lyse the cell membrane. The combination of lysozyme and a detergent is frequently used since this takes care of both the barriers. Lysozyme is also used in combination with osmotic shock or mechanical cell disruption methods. The main limitation of using lysozyme is its high cost. Other problems include the need for removing lysozyme from the product and the presence of other enzymes such as proteases in lysozyme samples.

4.9. Cell disruption using organic solvents

Organic solvents like acetone mainly act on the cell membrane by solubilizing its phospholipids and by denaturing its proteins. Some solvents like toluene are known to disrupt fungal cell walls. The limitations of using organic solvents are similar to those with detergents, i.e. the need to remove these from products and the denaturation of proteins. However, organic solvents on account of their volatility are easier to remove than detergents.

4.10. Cell disruption by osmotic shock

As discussed early in this chapter, osmotic pressure results from a difference in solute concentration across a semi permeable membrane. Cell membranes are semi permeable and suddenly transferring a cell from an isotonic medium to distilled water (which is hypotonic) would result is a rapid influx of water into the cell. This would then result in the rapid expansion in cell volume followed by its rupture, e.g. if red blood cells are suddenly introduced into water, these hemolyse, i.e. disrupt thereby releasing hemoglobin. Osmotic shock is mainly used to lyse mammalian cells. With bacterial and fungal cells, the cell walls need to be weakened before the application of an osmotic shock.

Osmotic shock is used to remove periplasmic substances (mainly proteins) from cells without physical cell disruption. In a large number of recombinant as well as non-recombinant gram negative bacteria, target proteins are secreted into the periplasmic space. If such cells are transferred to hypotonic buffers, the cells imbibe water through osmosis and the volume confined by the cell membrane increase significantly. The cell wall or capsule which is relatively rigid does not expand like the cell membrane and hence the material present in the periplasmic space is expelled out into the liquid medium (see Fig. 4.12).

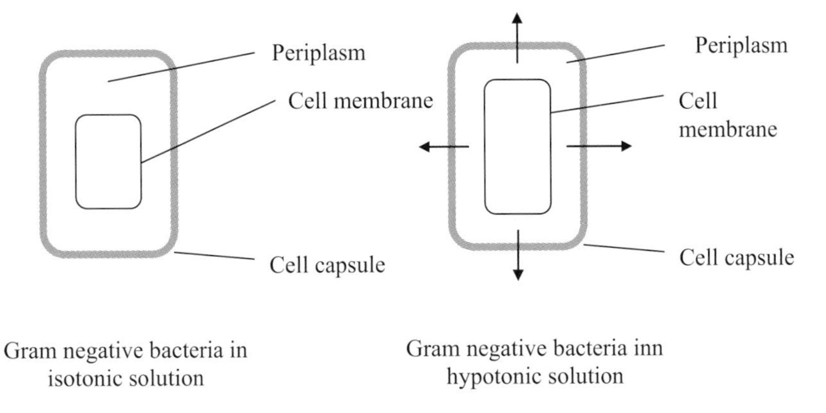

Fig. 4.12 Recovery of periplasmic products

Exercise problems

4.1. An intracellular antibiotic is being recovered by ultrasonication from 5 litres of bacterial cell suspension having a cell concentration of 15 g/l. Past experiences have shown that 50% of the antibiotic can be recovered in 40 minutes. Predict the time required for 90% recovery of the antibiotic.

4.2. A type of animal derived tissue is being ground in a bead mill to recover a pharmaceutical enzyme. It was estimated that 3000 joules of energy was used to reduce the average particle size of the tissue from 1 mm to 0.5 mm. Predict the amount of energy needed to further reduce the particle size to 0.1 mm. Assume that this grinding process follows Rittinger's law.

References

P.A. Belter, E.L. Cussler, W.-S. Hu, Bioseparations: Downstream Processing for Biotechnology, John Wiley and Sons, New York (1988).

R.L. Earle, Unit Operations in Food Processing, 2nd Edition, Pergamon Press, Oxford (1983).

M.R. Ladisch, Bioseparations Engineering: Principles, Practice and Economics, John Wiley and Sons, New York (2001).

P. Todd, S.R. Rudge, D.P. Petrides, R.G. Harrison, Bioseparations Science and Engineering, Oxford University Press, Oxford (2002).

Chapter 5

Precipitation

5.1. Introduction

Precipitation is an important traditional method for purifying proteins and nucleic acids. The most common example of precipitation based bioseparation is the Cohn fractionation method for purifying plasma proteins. Using this method, which consists of an array of precipitation steps, the various component proteins of human plasma are obtained in pure forms. Precipitation based bioseparation essentially involves selective conversion of a specific dissolved component of a complex mixture to an insoluble form using appropriate physical or physicochemical means. The insoluble form which is obtained as a precipitate (typically an easy to sediment solid) is subsequently separated from the dissolved components by appropriate solid-liquid separation techniques such as centrifugation (see Fig. 5.1). Crystallization is a special type of precipitation process where the solid is obtained in a crystalline form. Crystal formation takes place very slowly under highly controlled conditions and hence crystallization processes need to be carried out under highly optimized conditions. In contrast, a normal precipitation process does not need to be precisely controlled and the solid material obtained is amorphous in nature.

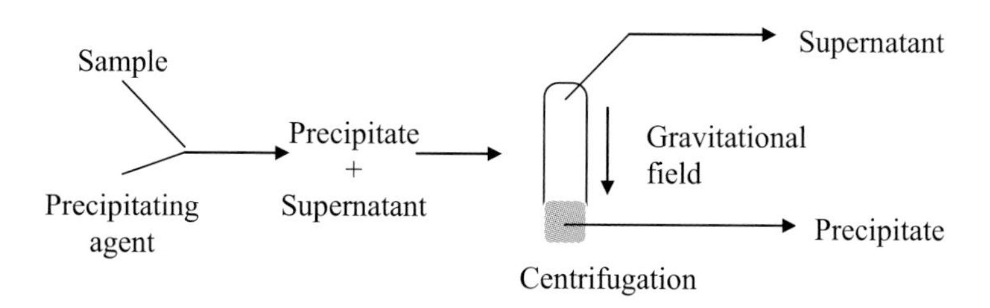

Fig. 5.1 Precipitation process

5.2. Factors utilized for precipitation

Biological macromolecules can be precipitated by:

1. Cooling
2. pH adjustment
3. Addition of solvents such as acetone and ethanol
4. Addition of anti-chaotropic salts such as ammonium sulphate and sodium sulfate
5. Addition of chaotropic salts such as urea and guanidine hydrochloride
6. Addition of biospecific reagents as in immunoprecipitation

The solubility of proteins in aqueous solutions depends on the temperature. Fig. 5.2 shows the solubility-temperature profiles of two proteins A and B. Quite clearly the solubility of A is more sensitive to the lowering of temperature than B. Hence, the separation of A and B could be carried out by lowering the temperature of the solution to the point where A is largely precipitated while B is still largely in solution. Cooling on its own is rarely used for precipitation. However cooling is an integral part of precipitation involving organic solvents and anti-chaotropic salts, partly due to the synergistic solubility lowering effect and partly due to the higher stability of biological macromolecules at lower temperatures.

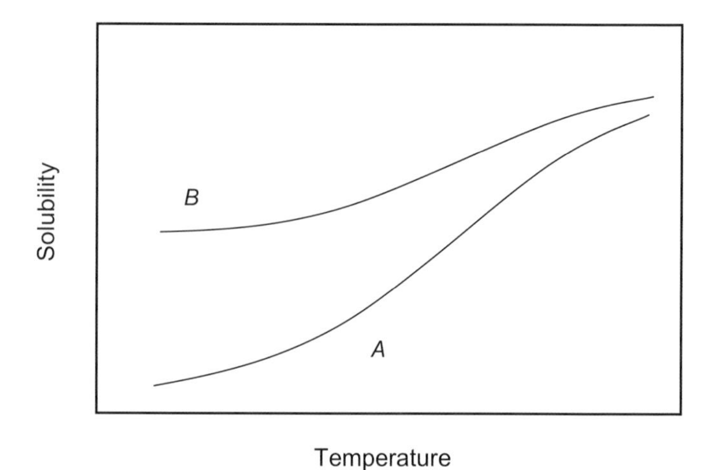

Fig. 5.2 Effect of temperature on protein solubility

Precipitation by using additives is governed by the thermodynamic equation:

$$\mu_{ppt} = \mu_{liq} \tag{5.1}$$

Where

μ_{ppt} = chemical potential of the precipitated solute
μ_{liq} = chemical potential of the dissolved solute

When a solute is being precipitated from its solution, the precipitate is mainly composed of the solute. So its chemical potential at a given temperature is constant. On the other hand the chemical potential of the dissolved solute depends on its concentration and is given by:

$$\mu_{liq} = \mu_{liq}(0) + RT \ln (x) \tag{5.2}$$

Where

$\mu_{liq}(0)$ = chemical potential of solute at standard reference state
x = solute concentration (mole-fraction)

The standard reference state chemical potential can be increased by adding substances such as salts and solvents. In the presence of such additives, the solute concentration in the solution phase must decrease in order that the chemical potential of the solute in solution be the same as that in the precipitate. This is evident from equations (5.1) and (5.2). The decrease in concentration takes place by precipitation.

The solubility of a protein depends on the solution pH, the minimum solubility being observed at its isoelectric point. Fig. 5.3 shows the pH-solubility curves for two proteins A and B. Separation of the two proteins could be carried out by maintaining the solution pH at pH_B (the isoelectric point of B). At this condition B would largely be precipitated while A would largely remain in solution. Separation would also be feasible at pH_A (the isoelectric point of A). At this condition A would largely be precipitated while B would largely remain in solution. As with cooling, pH adjustment on its own is rarely used for protein separation since once again the solubility differences are not large enough. However, the pH-solubility effect can be utilized in salt based protein precipitation processes for optimization of operating conditions.

Solvents such as ethanol and acetone precipitate proteins by decreasing the dielectric constant of the solution. The use of organic solvents in protein precipitation is widespread, an important example of this being plasma protein fractionation. It must be noted however that organic solvents denature proteins at room temperature. Hence organic solvent based protein precipitation processes are usually carried out at low temperatures in order to minimize denaturation. Organic solvents are also widely used for DNA and RNA precipitation.

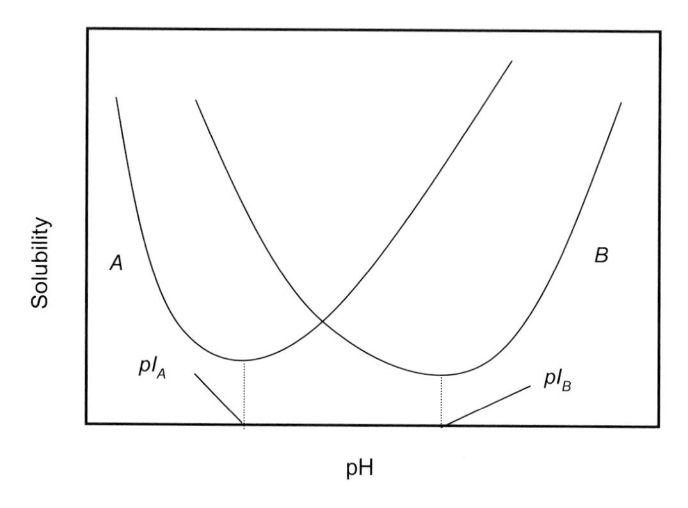

Fig. 5.3 Effect of pH on protein solubility

Anti-chaotropic salts such as ammonium sulphate and sodium sulphate expose hydrophobic patches on proteins by removing the highly structured water layer which usually covers these patches in solution. As a result hydrophobic residues on a protein molecule can interact with those on another and this eventually leads to aggregation and precipitate formation. Salts can also reduce the solubility of proteins by shielding charged groups which normally keep proteins apart in solution. When the electrostatic charge on protein molecules are shielded, the molecules can easily interact, form aggregates and eventually precipitate. The solubility of different proteins is reduced to different extents by salt addition. Based on this differential behaviour, separation of proteins is feasible, an important example being the partial purification of monoclonal antibody from cell culture supernatant. Fig. 5.4 shows the effect of ammonium sulphate concentration on the solubility of two proteins A and B. At low ammonium sulphate concentrations, increase in salt concentration results in increase in protein solubility. This is referred to as the *salting-in* effect. At higher ammonium sulphate concentrations, the solubility of proteins decrease very significantly with increase in salt concentration. This is referred to as the *salting-out* effect. Fig. 5.4 would suggest that it might be possible to separate A from B by adjusting the ammonium sulphate concentration such that B would largely be precipitated while A would largely remain in solution. The use of anti-chaotropic salts for protein purification is widespread. The three salts most commonly used are ammonium sulphate, sodium sulphate and sodium chloride. Anti-chaotropic salts exert a stabilizing influence on proteins, i.e. there is negligible denaturation. Anti-chaotropic salts could also be used in conjunction with other agents for DNA and RNA precipitation.

Chaotropic or chaos forming salts such as urea and guanidine hydrochloride precipitate proteins by denaturing them. These salts act by disrupting intra-molecular hydrogen bonds and hydrophobic interactions in proteins. They are not used for protein fractionation but are mainly used for protein refolding, particularly in the processing of inclusion bodies. Chaotropic salts are used in DNA and RNA purification since they precipitate proteins but leave nucleic acid molecules largely unaffected.

Immunoprecipitation relies on antigen-antibody recognition and binding. This in its simplest form can be carried out by treating an antigen containing solution with an antibody or vice-versa. When multivalent antigens react with antibodies in solution they form large molecular networks by cross-linking. These large complexes which eventually precipitate are called precipitins and these can easily be removed from solutions by standard solid-liquid separation techniques. Another way of carrying out immunoprecipitation is by treating an antigen with an insoluble form of its antibody or by treating an antibody with an insoluble form of its antigen.

5.3. Precipitation using organic solvents

Organic solvents precipitate proteins and other macromolecules by reducing the dielectric constant of the medium in which they are present. The governing equation for solvent based precipitation of proteins is:

$$\ln\left(\frac{S}{S_w}\right)=\left(\frac{A}{RT}\right)\left(\left(\frac{1}{e_w}\right)-\left(\frac{1}{e}\right)\right) \tag{5.3}$$

Where

S = solubility of the protein (kg-moles/m^3)

S_w = solubility of protein in water (kg-moles/m^3)

A = a constant

e = dielectric constant of the medium (-)
e_w = dielectric constant of water (-)

The dielectric constant of water at 25 degrees centigrade is 78.3 while that of ethanol at the same temperature is 24.3. Thus equation (5.3) would suggest that a protein would have a lower solubility in an ethanol-water mixture than in water itself. Higher concentrations of organic solvents can denature proteins. Organic solvents usually bind to specific locations on the protein molecules and thus disrupt the hydrophobic interactions which hold the protein structure in place. Hence very small amounts of organic solvent are used in precipitation processes and these are carried out at low temperatures to minimize denaturation. Acetone and aliphatic alcohols such as methanol, ethanol, propanol and butanol are mainly used for protein precipitation. The extent of protein denaturation by aliphatic alcohols increases with increase in chain length, e.g. butanol causes more denaturation than ethanol.

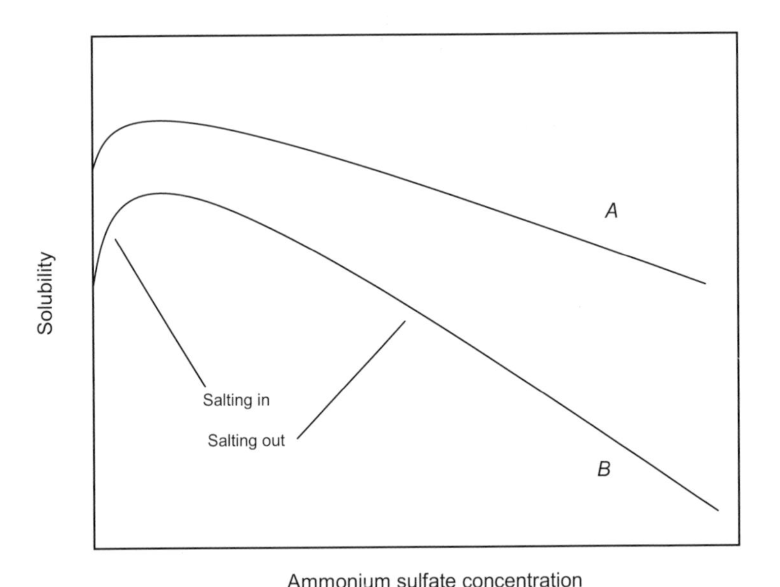

Fig. 5.4 Protein precipitation using anti-chaotropic salt

Nucleic acids such as DNA and RNA can also be precipitated using organic solvents such as ethanol. Nucleic acid precipitation processes are carried out at very low temperatures, typically -20 degrees centigrade.

The precipitation of nucleic acids by organic solvents can also be described using equation (5.3).

A major application of organic solvent based precipitation in bioseparation is the Cohn fractionation method for human plasma protein purification. In this method a series of ethyl alcohol precipitation steps are carried out using different alcohol concentrations, solution pH values and precipitation temperatures. Using this method a series of products such as serum albumin, immunoglobulin G, plasminogen, lipoproteins and prothrombin can be obtained.

Example 5.1

The solubility of ovalbumin in water is 390 kg/m^3. When 30 ml of ethanol was added to 100 ml of a 50 mg/ml ovalbumin solution in water, 33% of the protein was found to precipitate. How much protein would precipitate if 100 ml of ethanol were added to 100 ml of a similar protein solution at the same temperature? Assume that the dielectric constant of the medium varies linearly with volumetric composition of the two solvents.

Solution

As already mentioned in the text (after equation (5.3)), the dielectric constant of water is 78.3 and that of pure ethanol is 24.3. If the dielectric constant varies linearly with volumetric composition in the intermediate range, the dielectric constant of the mixture obtained by 30 ml of ethanol with 100 ml of aqueous solution is 65.8. When 33% of the ovalbumin is precipitated, the concentration in the solution is 33.5 mg/ml. Using equation (5.3) we get:

$$\ln\left(\frac{33.5}{390}\right)=\left(\frac{A}{RT}\right)\left(\left(\frac{1}{78.3}\right)-\left(\frac{1}{65.8}\right)\right) \qquad (5.a)$$

The dielectric constant of the solution obtained by mixing 100 ml of ethanol with 100 ml of aqueous solution is 51.3. Using equation (5.3) we get:

$$\ln\left(\frac{S}{390}\right)=\left(\frac{A}{RT}\right)\left(\left(\frac{1}{78.3}\right)-\left(\frac{1}{51.3}\right)\right) \qquad (5.b)$$

From equations (a) and (b) we get:

$S = 0.204$ mg/ml

Therefore percentage of protein precipitated is 99.6%.

5.4. Precipitation using anti-chaotropic salts

Salts such ammonium sulphate and sodium chloride are referred to as anti-chaotropic or kosmotropic salts. The solubility of proteins in the presence of such salts depend on the salt concentration as shown in Fig. 5.4. The *salting in* effect, i.e. increase in protein solubility with increase in salt concentration can be explained in terms of these salts providing a distinct electrostatic double-layer surrounding the proteins. The presence of the double layer provides the stability needed to keep the protein molecules in solution and hence increases their solubility. At higher salt concentrations, the electrostatic double-layers become diffuse (i.e. indistinct) and at such conditions, the salts can actually shield the electrostatic charges present on the proteins. This reduces the protein-protein electrostatic repulsion forces leading to protein aggregation and hence reduction in solubility. However, the main reason for the reduced protein solubility at higher salt concentrations is the increase in protein-protein hydrophobic interaction. Also the association of large amounts of water molecules with the salt molecules results in less water molecules being available for dissolving proteins.

Salt induced protein precipitation is governed by the Cohn equation, a form of which is shown below:

$$\ln(S) = B - K_s C_s \tag{5.4}$$

Where

B = constant

K_s = salting out constant (m^3/kg-moles)

C_s = salt concentration (kg-moles/m^3)

The constant B is the natural log of the theoretical solubility of the protein in salt free water. The Cohn equation is valid only in the salting-out region of the precipitation process. The constant B depends on the protein, the temperature and the solution pH. K_s is independent of the temperature and pH but depends on the salt and the protein. Salt induced precipitation is usually carried out at low temperatures typically 4 degrees centigrade. This is primarily due to the fact that the solubility of proteins decreases with drop in temperatures and hence more protein can be precipitated by such synergistic action. The second reason is the enhancement in protein stability at lower temperatures.

When using ammonium sulphate as a precipitating agent, two operational approaches could be used:

1. Direct addition of solid ammonium sulphate crystals to the sample
2. Addition of a saturated ammonium sulphate solution to sample

The first approach is preferred in small scale applications. However, this approach suffers from the unevenness of ammonium sulphate concentration in the precipitating medium caused due to the salt dissolution process. This may lead to localized higher salt concentration zones and result in uneven precipitate formation. In most protein precipitation processes, the second approach is preferred.

A terminology that is commonly used in ammonium sulphate precipitation is the *ammonium sulphate-cut*. A 50% ammonium sulphate cut refers to the use of an ammonium sulphate concentration in the precipitating medium which corresponding to 50% of the saturation concentration for this salt. In many purification processes, differential ammonium sulphate cuts are commonly used, i.e. the supernatant from one precipitation step is subjected to further ammonium sulphate precipitation. For example, in a 30-50% ammonium sulfate cut, the sample is first subjected to precipitation using 30% salt saturation. The supernatant thus obtained in then subjected to precipitation using 50% salt saturation. In many cases, the precipitates are re-dissolved in buffer followed by further salt precipitation.

Example 5.2

Ammonium sulfate is being used to precipitate a humanized monoclonal antibody from 10 litres of cell culture media, the initial concentration of the antibody in this liquid being 0.5 mg/ml. Solid ammonium sulfate is added to the liquid such that the concentration of the salt is 1.5 kg-moles/m^3. This results in the precipitation of 90% of the antibody. When the ammonium sulfate concentration of the mixture is raised to 1.75 kg-mole/m^3, a further 76.5% of the remaining antibody is precipitated. Predict the ammonium sulfate concentration needed for total antibody precipitation. What is the solubility of the antibody in ammonium sulfate free aqueous medium?

Solution

The antibody concentration in solution after 1.5 M ammonium sulfate precipitation is 0.05 mg/ml = 3.2×10^{-7} kg-mole/m^3

Similarly the antibody concentration in solution after 1.75 M ammonium sulfate precipitation is 0.0118 mg/ml = 7.58×10^{-8} kg-mole/m^3.

Using equation (5.4) we get:

$$\ln\left(3.2\times10^{-7}\right) = B - K_S \times 1.5$$

and

$$\ln\left(7.58\times10^{-8}\right) = B - K_S \times 1.75$$

Therefore:

B = -6.26

K_S = 5.8 m^3/kg-mole

For total precipitation we have to assume a very small value of S since equation (5.4) becomes undefined at S = 0. Assuming that $S = 6 \times 10^{-11}$ kg-mole/m^3 and using equation (5.4), we get:

C_S = 2.97 kg-mole/m^3 = 2.97 M

B = ln (solubility)

Therefore the antibody solubility = 1.91×10^{-3} kg-mole/m^3 = 296 kg/m^3.

5.5. Mechanism of precipitate formation

Precipitate formation is a time dependent process which basically involves the formation of tiny particles due to association of macromolecules. These particles may then collide with each other and further increase in size. If we track a precipitation process by measuring the turbidity of the medium in which the precipitation is taking place as a function of time, we typically get a plot similar to that shown in Fig. 5.5. The turbidity of a suspension is a measure of its suspended solids content and this value is analogous to the absorbance term discussed in chapter 2 of this book. The turbidity is calculated by measuring the incident and transmitted light of a given wavelength through a sample held within an optical cell (also called a cuvette). The greater the value of turbidity, the greater is the suspended solids content.

A precipitation process has the following stages:

1. Mixing
2. Nucleation
3. Diffusion limited growth
4. Convection limited growth

The first step involves the formation of a homogeneous mixture of the various components involved in precipitation, i.e. the macromolecules, the solvent and the precipitating agent. Mixing is not an instantaneous process. It usually takes place over a finite amount of time, the amount of time depending on the properties of the components as well as on the processing conditions, e.g. mixing intensity and temperature. The time needed for mixing is given by:

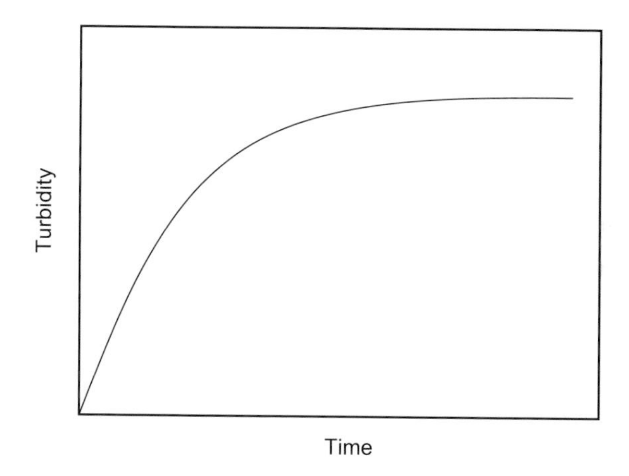

Fig. 5.5 Change in turbidity during precipitation process

$$\theta = \left(\frac{l_e^2}{4D} \right) \tag{5.5}$$

Where

θ = mixing time (s)

l_e = average eddy length (m)

D = diffusivity (m^2/s)

The average eddy length scale depends on the volume of the system, the density and viscosity of the medium, as well as on the power input for mixing.

The nucleation step involves the formation of very minute particles. The particles initially formed in protein precipitation are multimers, i.e. aggregates of three or more protein molecules. Nucleation usually initiates at regions having localised supersaturation. The supersaturation results from uneven mixing. The extent of supersaturation determines the nature of the precipitates: very high degrees of supersaturation result in formation of gelatinous precipitates which are difficult to process by filtration or membrane separation. Controlled supersaturation produces amorphous precipitates which are both easy to filter and centrifuge.

The deposition of more proteins on the small nucleation particles takes place by a diffusion controlled mechanism and this leads to the formation of bigger particles, typically ranging from less than a micron to several microns in size. In this step, the rate of particle formation is dependent on the physicochemical properties of the protein and the liquid medium alone. The rate of formation of particles of a particular size is governed by second order kinetics:

$$\frac{dn}{dt} = -\kappa n^2 \tag{5.6}$$

Where

n = number of particles of a given size (-)

κ = rate constant

The rate constant κ depends on the diffusivity and diameter of the protein.

The particles produced by diffusion limited growth further increase in size by colliding with one another, thereby absorbing the momentum and forming bigger particles. These particles are typically a few to several hundred microns in size. This part of the precipitation process is convection dependent since the frequency of collision is a function of the hydrodynamic conditions in the system. The greater the extent of mixing, the greater is the frequency of collision. The rate of precipitate formation of a particular size in convection limited growth is given by:

$$\frac{dn}{dt} = -\varphi n^2 \tag{5.7}$$

Where

φ = rate constant

The rate constant φ depends on the size and sticking tendency of the particles, the volume of the system, the density and viscosity of the medium in addition to the power input for mixing. It must be noted however, that the collision of particles would in some cases result in their breakage. Indeed, towards the end of a precipitation process, the rate of particle formation is almost matched by the rate of particle breakage, resulting in no further particle formation. When a precipitate is stored for some time without mixing, a process called ageing, some changes in particle size and size distribution takes place. This ageing process facilitates separation of precipitates from the supernatant. Typical ageing duration for ammonium sulphate precipitated proteins is 6 to 12 hours at 4 degrees centigrade.

Exercise problems

5.1. Monoclonal antibody (MW = 155,000 kg-moles/m^3) is being purified from hybridoma cell culture supernatant by ammonium sulfate precipitation. The antibody concentration in the starting material is 0.5 g/l and 30% of the antibody could be precipitated using 1 M ammonium sulfate cut. What salt concentration would have resulted in 90% precipitation of the monoclonal antibody? The solubility of the antibody in salt free water is 390 g/l.

5.2. Predict the solubility of a protein having a molecular weight of 67,000 kg/kg-mole in water if its solubility values in 1.5 M and 2 M ammonium sulfate solutions are 18 g/l and 6 g/l respectively.

5.3. Murine monoclonal antibody is to be recovered from 100 litres of cell culture supernatant (having antibody concentration of 0.2 kg/m^3) by ammonium sulphate precipitation. The molecular weight of the antibody is 155,000 kg/kg-mole while that of ammonium sulphate is 132 kg/kg-mole. How many kilograms of ammonium sulphate would be needed to precipitate 95% of the antibody? A preliminary experiment carried out using 1 litre of a similar cell culture supernatant showed that 180 mg of the antibody could be precipitated at 1.75 M ammonium sulphate concentration. The solubility of the antibody in salt free water is 390 kg/m^3.

References

B. Atkinson, F. Mavituna, Biochemical Engineering and Biotechnology Handbook, Macmillan Publishers Ltd., Surrey (1983) p. 1.

J.E. Bailey, D.F. Ollis, Biochemical engineering Fundamentals, 2nd edition, McGraw Hill, New York (1986).

D.J. Bell, M. Hoare, P. Dunnill, Formation of Protein Precipitates and their Centrifugal Recovery, in: A. Fiechter (Ed.), Advances in Biochemical Engineering vol. 26, Springer-Verlag, Berlin (1983).

P.A. Belter, E.L. Cussler, W.-S. Hu, Bioseparations: Downstream Processing for Biotechnology, John Wiley and Sons, New York (1988).

P.R. Foster, Blood Fractionation, in: M. Howe-Grant (Ed.), Kirk Othmer Encyclopedia of Chemical Technology, Wiley, New York, (1994) p. 977.

M.R. Ladisch, Bioseparations Engineering: Principles, Practice and Economics, John Wiley and Sons, New York (2001).

R.K. Scopes, Protein Purification: Principle and Practice, Springer-Verlag, New York (1982).

S. Doonan (Ed.), Protein Purification Protocols, Humana Press, New Jersey (1996).

Chapter 6

Centrifugation

6.1. Introduction

A centrifuge is a device that separates particles from suspensions or even macromolecules from solutions according to their size, shape and density by subjecting these dispersed systems to artificially induced gravitational fields. Centrifugation can only be used when the dispersed material is denser than the medium in which they are dispersed. Table 6.1 lists the densities of different biological substances that are usually separated by centrifugation.

Table 6.1 Densities of biological material

Material	Density (g/cm^3)
Microbial cells	1.05 – 1.15
Mammalian cells	1.04 – 1.10
Organelles	1.10 – 1.60
Proteins	1.30
DNA	1.70
RNA	2.00

Based on data shown in Table 6.1 one may wrongly assume that proteins and nucleic acids would settle faster than cells and organelles. Biological macromolecules in aqueous solution exist in an extensively hydrated form i.e. in association with a large number of water molecules. Hence the effective densities of these substances in solution are only slightly higher than that of water. Also as mentioned in chapter 2, these macromolecules are significantly smaller than cells. The substances listed in Table 6.1 would settle at extremely low velocities under gravity and hence separation would not be feasible. In a centrifugation process, these settling rates are amplified using an artificially induced gravitational field. Cells, sub cellular components, virus particles and precipitated forms of proteins and nucleic acids are easy to separate by centrifugation. When macromolecules such as proteins, nucleic acids and carbohydrates need to be separated, normal centrifuges cannot be used and special devices called ultracentrifuges which generate very strong artificial gravitational fields are used. The principle of separation by centrifugation is shown in Fig. 6.1.

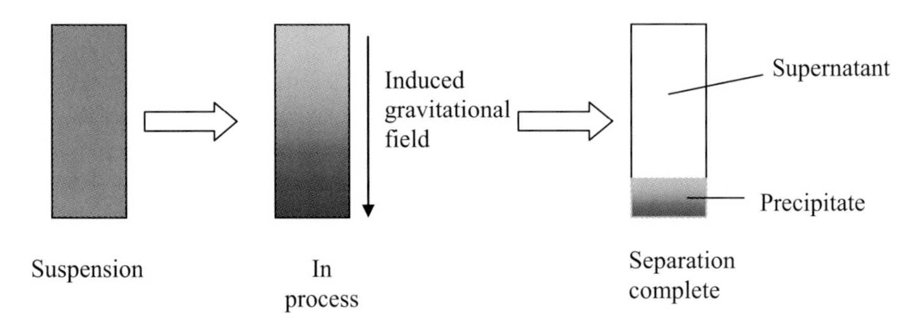

Fig. 6.1 Centrifugation

Centrifuges are classified into two categories:

1. Laboratory centrifuges
2. Preparative centrifuges

6.2. Laboratory centrifuge

Laboratory centrifuges are used for small-scale separation and clarification (i.e. removal of particles from liquids). Typical liquid volumes handled by such devices are in the range of 1 – 5000 ml. Fig. 6.2 shows a diagram of a laboratory centrifuge. The material to be centrifuged is distributed into appropriate numbers of centrifuge tubes (which look like test tubes) which are in turn attached in a symmetric manner to a rotating block called the rotor. There are two types of rotors: fixed angle rotors and swing out rotors. A fixed angle rotor holds the centrifuge in a fixed manner at particular angle to the axis of rotation. Swing out rotors hold the tubes parallel to the axis of rotation while the rotor is stationary but when the rotor is in motion, the tubes swing out such that they are aligned perpendicular to the axis of rotation. When the centrifuge tubes are spun, the centrifugal action creates an induced gravitational field in an outward direction relative to the axis of rotation and this drives the particles or precipitated matter towards the bottom of the tube. Typical rotation speeds of laboratory centrifuges range from 1,000 – 15,000 rpm. The magnitude of the induced gravitational field is measured in terms of the *G* value: a *G* value of 1000 refers to an induced field that is thousand time stronger than that due to gravity. The *G* value which is also referred to as the *RCF* (relative centrifugal force) value depends on the rotation speed as well as the manner in which the centrifuge tubes are held by the rotor:

$$G = \frac{r\omega^2}{g} = \frac{2\pi r n^2}{g} \tag{6.1}$$

Where

r = distance from the axis of rotation (m)
ω = angular velocity (radians/s)
g = acceleration due to gravity (m/s^2)
n = rotation speed (/s)

Quite clearly the *G* value in a centrifuge tube will depend on the location, the highest value being at the bottom of the tube and the lowest value being at the top. This implies that a particle will experience increasing *G* values while moving towards the bottom of the centrifuge tube. In most cases the average *G* value that a particle is likely to experience i.e. the numerical mean of the maximum and minimum values is used for process calculations. Nomograms provided by centrifuge manufacturers correlating the radial distance and rotation speed with the *G* value are commonly used in the laboratory-scale calculations. Typical *G* values employed in laboratory centrifuges range from 1,000 – 20,000.

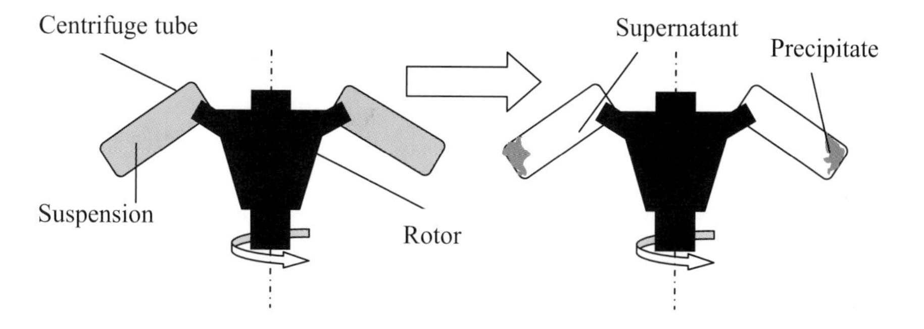

Fig. 6.2 Laboratory centrifugation

It is virtually impossible to make very exact calculations for a laboratory centrifugation process. This is due to the fact that it usually takes a certain amount of time after start-up for the rotation speed of the centrifuge to reach the operating value. Similarly it takes a certain amount of time for the rotation speed to decrease from the operating speed to zero at the end of the process. Moreover as mentioned previously, the settling particles go through different G value zones while moving toward the bottom of the centrifuge tube. An empirical correlation is commonly used for estimating the complete centrifugation time:

$$t = \frac{k}{S} \tag{6.2}$$

Where
k = k-factor of the centrifuge
S = Svedberg coefficient of the material being precipitated
t = complete sedimentation time (min)

One Svedberg unit (S) is equal to $10^{-13} s$ (s is defined in chapter 2). A smaller k-factor results in a faster centrifugation process. The k-factor can be calculated using the empirical correlation shown below:

$$k = 2.53 \times 10^{11} \left(\frac{\ln(r_{max} - r_{min})}{rpm_{max}^2} \right) \tag{6.3}$$

Where
r_{max} = radial distance from the axis to the bottom of the tube (cm)
r_{min} = radial distance from the axis to the top of the tube (cm)
rpm_{max} = maximum rotation speed (/min)

For the same rotation speed, a fixed angled rotor would have a lower k-factor on account of the smaller difference between r_{max} and r_{min} (see Fig. 6.3). Hence the time required for precipitating a given sample would be less with the fixed angled rotor. This is intuitively evident from looking at Fig. 6.3 which shows that the distance the particles have to travel for precipitation is less with the angled rotor. However, fixed angled rotors are heavier and require much higher energy to operate Swing out rotors are preferred for centrifuging substances with high sedimentation coefficients such as cells and coarse particles while precipitated macromolecules and finer particles are centrifuged using fixed angled rotors.

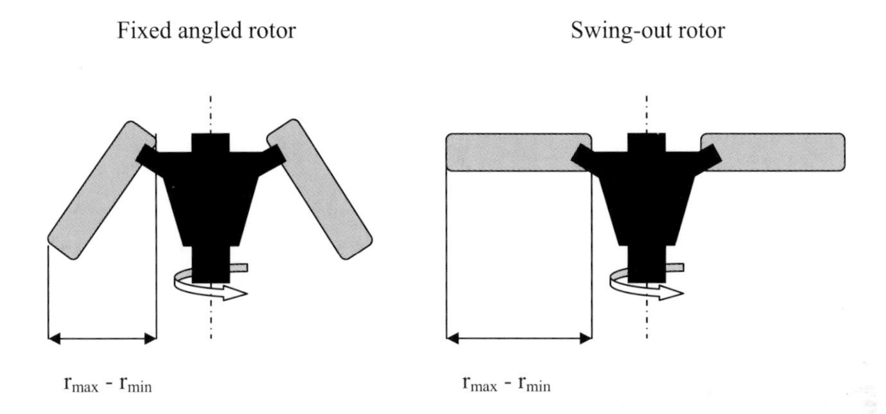

Fig. 6.3 Fixed angled and swing-out rotors

6.3. Preparative centrifuge

Preparative centrifuges can handle significantly larger liquid volumes than laboratory centrifuges, typically ranging from 1 litre to several thousand litres. Preparative centrifuges come in a range of designs, the common feature in these being a tubular rotating chamber. The suspension to be centrifuged is fed into such a device from one end while the supernatant and precipitate are collected from the other end of the device in a continuous or semi-continuous manner. A simple diagram of the most common type of preparative centrifuge (the tubular bowl centrifuge) is shown in Fig. 6.4. Typical rotating speeds for preparative centrifuges range from 500 - 2000 rpm.

The motion of a particle at any point within a tubular centrifuge in the radial direction is governed by the following force balance equation:

$$\left[\frac{4}{3}\pi\left(\frac{d}{2}\right)^3(\rho_S - \rho_L)\right]\left[r\omega^2\right] = 6\pi\mu\left(\frac{d}{2}v\right) \qquad (6.4)$$

Where

d = diameter of the particle (m)
ρ_S = density of the particle (kg/m^3)
ρ_L = density of the liquid (kg/m^3)
r = distance of the particle from the axis of rotation (m)
ω = angular velocity (radians/s)
μ = viscosity of the liquid (kg/m s)
v = velocity of the particle (m/s)

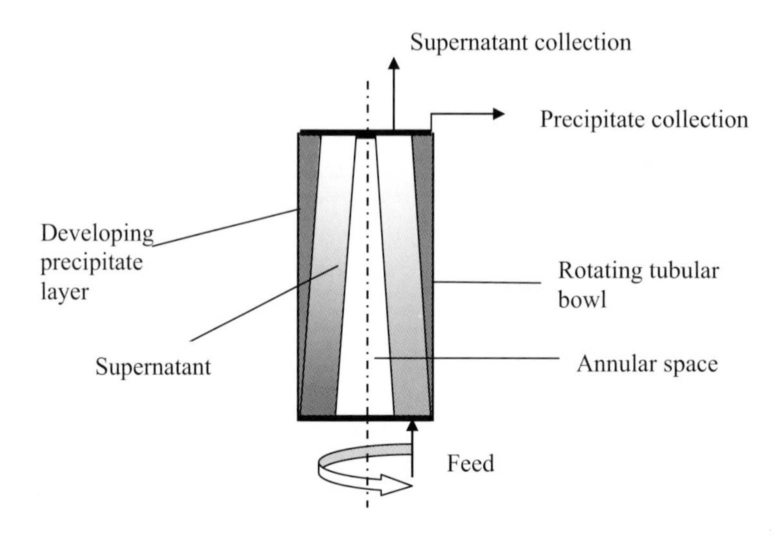

Fig. 6.4 Solid-liquid separation using a tubular bowl centrifuge

The term on the left hand side of equation (6.4) is the buoyancy term while the term on the right hand side is the Stokes drag term. In this equation we assume the particle to be moving at constant velocity in the radial direction. Rearranging equation (6.4):

$$v = \frac{d^2(\rho_S - \rho_L)r\omega^2}{18\mu} \tag{6.5}$$

The velocity in the radial direction can be replaced by (dr/dt). Hence:

$$\frac{dr}{dt} = \frac{d^2(\rho_S - \rho_L)\omega^2 r}{18\mu} \tag{6.6}$$

Integrating equation (6.6), putting in appropriate limits we get:

$$\int_{r_1}^{r_2} \frac{dr}{r} = \int_0^t \frac{d^2(\rho_S - \rho_L)\omega^2}{18\mu} dt$$

Therefore:

$$\ln\left(\frac{r_2}{r_1}\right) = \frac{d^2(\rho_S - \rho_L)\omega^2 t}{18\mu} \tag{6.7}$$

Equation (6.7) describes the motion of a particle in the radial direction within a centrifuge from radial location r_1 to r_2. In a preparative centrifuge, the motion of a particle takes place in two directions: radial (due to centrifugation) and axial (due to flow of feed). The flow of the feed within the tubular centrifuge is annular in nature, i.e. there is an empty cylindrical shell near the axis of rotation. This is due to the fact that the liquid within the centrifuge is forced towards the wall of the tubular bowl due to centrifugal force. The flow of feed within a tubular bowl centrifuge is shown in Fig 6.5: the diameter of the tubular bowl is r_t while the diameter of the annular space is r_a. The flow of feed within the centrifuge is in the axial direction (along z-axis).

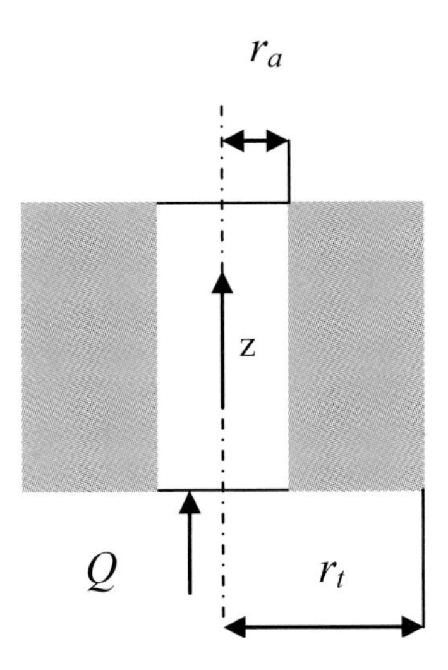

Fig. 6.5 Annular flow in tubular bowl centrifuge

The velocity of a particle along the z axis is given by:

$$\frac{dz}{dt} = \frac{Q}{\pi\left(r_t^2 - r_a^2\right)} \tag{6.8}$$

Where

Q = feed flow rate (m^3/s)

Dividing equations (6.6) by equation (6.8) and integrating we get:

$$\int_{r_a}^{r_t} \frac{dr}{r} = \int_0^z \frac{\pi d^2 \left(\rho_S - \rho_L\right) \omega^2 \left(r_t^2 - r_a^2\right)}{18 \mu Q} dz \tag{6.9}$$

Therefore:

$$\ln\left(\frac{r_t}{r_a}\right) = \frac{\pi d^2 (\rho_S - \rho_L)\omega^2 (r_t^2 - r_a^2) z}{18\mu Q} \tag{6.10}$$

Rearranging, we get:

$$Q = \left[\frac{d^2(\rho_S - \rho_L)}{18\mu}\right]\left[\frac{\pi z (r_t^2 - r_a^2)\omega^2}{\ln\left(\frac{r_t}{r_a}\right)}\right] \tag{6.11}$$

This equation can be written as:

$$Q = \left[\frac{d^2(\rho_S - \rho_L) g}{18\mu}\right]\left[\frac{\pi z (r_t^2 - r_a^2)\omega^2}{g \ln\left(\frac{r_t}{r_a}\right)}\right] \tag{6.12}$$

Where

g = acceleration due to gravity (m/s^2)

The first term in the square brackets in equation (6.12) is referred to as sedimentation velocity, i.e. the velocity of a particle settling due to gravity alone and is denoted by v_g. The second term in the square brackets is referred to as the sigma factor (Σ) of the centrifuge. Therefore the feed flow rate into the tubular centrifuge can be expressed in terms of the product of v_g and Σ, the first representing properties of the material being handled and the second representing the those of the equipment.

$$Q = v_g \Sigma \tag{6.13}$$

Where

v_g = velocity due to gravity (m/s)

Σ = sigma factor of the centrifuge (m^2)

A disc stack centrifuge is a special type of preparative centrifuge which is compact in design and gives better solid-liquid separation than the standard tubular bowl centrifuge. Fig. 6.6 shows the working principle of a disc stack centrifuge. The feed enters from the top of the device and is distributed at the bottom of the disk bowl through a hollow drive shaft. The particles are thrown outward and these come into contact with the angled disc stack. Once this happens they slide down the disc, are collected at the periphery of the bowl and discharged from the device in the form of a slurry. The liquid flows up the device along the central regions and is discharged from the top.

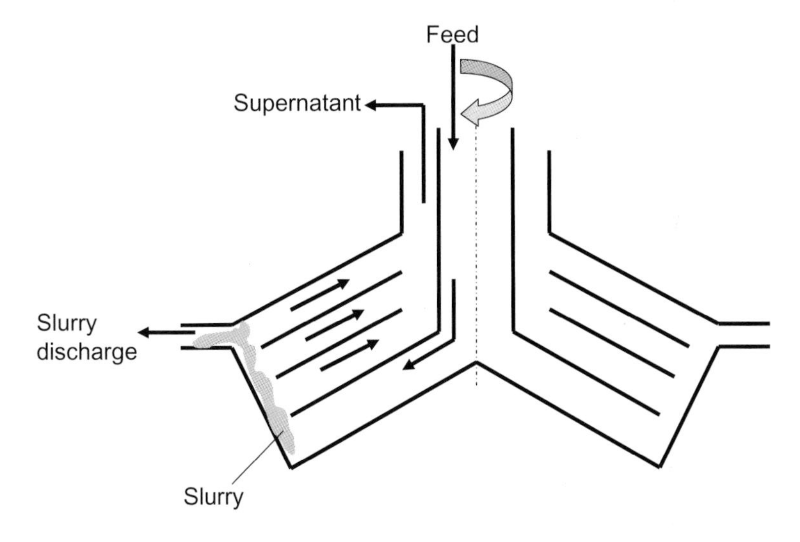

Fig. 6.6 Disk stack centrifugation process

6.4. Ultracentrifugation

An ultracentrifuge is a special type of centrifuge in which the rotor rotates at a much higher speed than a standard centrifuge. Typical rotation speeds in ultracentrifuges range from 30000 rpm – 50000 rpm. An ultracentrifuge is usually used for separating macromolecules from solvents or for fractionating mixtures of macromolecules. Ultracentrifuges are used for analytical as well as for preparative applications. An analytical ultracentrifuge (AUC) is mainly used for studying the properties of macromolecules as well as for analyzing complex mixtures of macromolecules. Preparative ultracentrifuges are used to purify macromolecules such as proteins and nucleic acids based on their physical properties such as size, molecular weight, density and mobility. The high rotating speeds used in ultracentrifuges can generate considerable amount of heat. Therefore cooling arrangements are required in these devices.

Exercise problems

6.1. Bacterial cells were centrifuged using 5 cm long centrifuge tubes in a laboratory centrifuge having an angled rotor which held the tubes at 60 degree angle with the axis of rotation. The top of the tubes were 5 cm away from the axis and it took 15 minutes at a rotation speed of 10,000 rpm to completely sediment the cells. Calculate the sedimentation coefficient of the cells.

6.2. A yeast suspension is to be clarified using a tubular bowl centrifuge at a flow rate of 1.5 l/min at 4 degrees centigrade. The yeast cells are 5 microns in diameter and have a density of 1.15 g/cm^3. Assuming that the viscosity of the liquid medium is the same as that of water, calculate the Σ factor required.

References

B. Atkinson, F. Mavituna, Biochemical Engineering and Biotechnology Handbook, Macmillan Publishers Ltd., Surrey (1983) p. 1.

J.E. Bailey, D.F. Ollis, Biochemical engineering Fundamentals, 2nd edition, McGraw Hill, New York (1986).

P.A. Belter, E.L. Cussler, W.-S. Hu, Bioseparations: Downstream Processing for Biotechnology, John Wiley and Sons, New York (1988).

C. J. Geankoplis, Transport Processes and Separation Process Principles, 4th edition, Prentice Hall, Upper Saddle River (2003).

M.R. Ladisch, Bioseparations Engineering: Principles, Practice and Economics, John Wiley and Sons, New York (2001).

W.L. McCabe, J.C. Smith, P. Harriott, Unit Operations of Chemical Engineering, 7th edition, McGraw Hill, New York (2005).

Chapter 7

Extraction

7.1. Introduction

An extraction process makes use of the partitioning of a solute between two immiscible or partially miscible phases. For example the antibiotic penicillin is more soluble in the organic solvent methyl isobutyl ketone (MIBK) than in its aqueous fermentation medium at acidic pH values. This phenomenon is utilized for penicillin purification. When the extraction takes place from one liquid medium to another, the process is referred to as liquid-liquid extraction. When a liquid is used to extract solutes from a solid material, the process is referred to as solid-liquid extraction or leaching. In this chapter we will mainly discuss liquid-liquid extraction. When a supercritical fluid is used as an extracting solvent, the process is referred to as supercritical fluid extraction (SFE). SFE will be briefly discussed at the end of this chapter.

Typical applications of extraction in bioprocessing include:

1. Purification of antibiotics
2. Purification of alkaloids
3. Protein purification using aqueous two-phase systems
4. Purification of peptides and small proteins
5. Purification of lipids
6. Purification of DNA

7.2. Solvent systems

Solvent extraction as used in the bio-industry can be classified into three types depending on the solvent systems used:

1. Aqueous/non-aqueous extraction
2. Aqueous two-phase extraction
3. Supercritical fluid extraction

Low and intermediate molecular weight compounds such as antibiotics, alkaloids, steroids and small peptides are generally extracted using aqueous/non-aqueous solvent systems. Biological macromolecules such as proteins and nucleic acids can be extracted by aqueous two-phase systems. Supercritical fluids are used as extracting solvents where organic or aqueous solvents cannot be used satisfactorily.

Ideally, the two solvents involved in an extraction process should be immiscible. However, in some extraction processes partially miscible solvent systems have to be used. For partially miscible solvent systems, particularly where the solute concentration in the system is high, triangular or ternary phase diagrams such as shown in Fig. 7.1 are used. In such diagrams the concentration of the components are usually expressed in mole fraction or mass fraction. Fig. 7.1 shows the phase diagram for a solute A, its initial solvent B and its extracting solvent C. Such phase diagrams rely on the fact that all possible composition of the three components can be represented by the area within the triangle. The composition of the mixture represented by point H on the diagram is such that content of A is proportional to HL, content of B is proportional to HJ and content of C is proportional to HK. The curve shown in Fig. 7.1 is called the binodal solubility curve. The area under the curve represents the two-phase region. Any mixture represented by a point within this region will split up into two phases in equilibrium with each other. For a mixture having an overall composition H, the composition of the two phases are represented by points P and Q which are obtained from the points of intersection of the binodal solubility curve with the *tie line* PQ which passes through H. The *tie lines* are straight lines which connect together the compositions of the two phases, which are in equilibrium with one another. These *tie lines* are experimentally determined. The point F on the binodal solubility curve is called the *plait point*. The area above the binodal solubility curve represents the single-phase region where all three components in the system are mutually miscible.

7.3. Theory of extraction

The different components involved in an extraction processes are summarized in Fig. 7.2. The feed consists of the solute to be extracted in its original solvent e.g. penicillin in fermentation media. The extracting solvent (e.g. MIBK) is the phase to which the solute (i.e. penicillin in this

Liquid-SF extraction

The efficient dissolving power of a supercritical fluid means that the solute of interest and the initial solvent are both likely to be soluble in it. Ternary phase diagrams similar to those used in liquid-liquid extraction using partially miscible liquids are also used in liquid-SF extraction. Fig. 7.20 shows the ternary phase diagram for ethanol-water-supercritical carbon dioxide system. As evident from the phase diagram, supercritical carbon dioxide is not suitable for breaking water-ethanol azeotrope. However, is useful for removing ethanol from ethanol water mixture, an application of this being the manufacture of alcohol free beer.

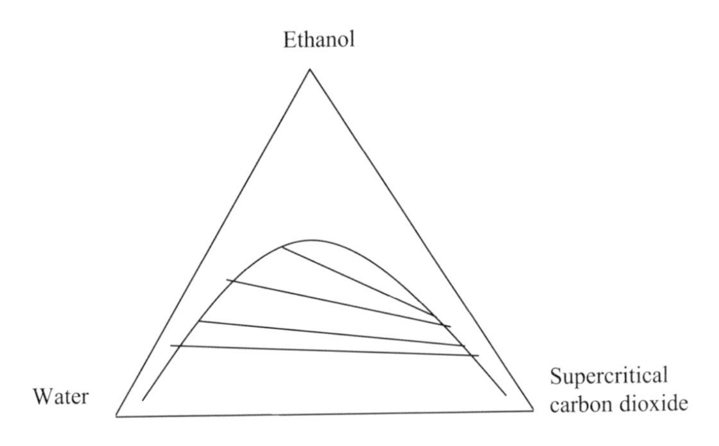

Fig. 7.20 Ethanol-water-supercritical carbon dioxide ternary phase diagram

Fig. 7.21 shows the set-up used for obtaining relatively concentrated ethanol from a dilute ethanol-water mixture. Gaseous carbon dioxide is first converted to a supercritical fluid by adiabatic compression, which is then fed into the extraction vessel. The ethanol-water mixture is fed into this vessel at the operating pressure in a counter-current direction. The supercritical fluid preferentially extracts the ethanol and leaves the extractor from the top while the raffinate (predominantly water) exits from the bottom of the extractor. The extract is sent through a pressure reduction valve into an expansion chamber where gaseous carbon dioxide and liquid ethanol-water mixture is obtained. The carbon dioxide is separated from the ethanol-water mixture and sent back to the compressor.

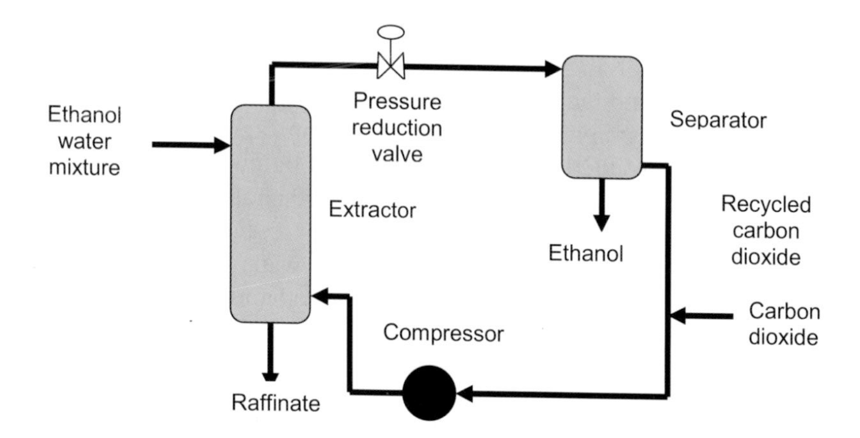

Fig. 7.21 Enrichment of ethanol using supercritical fluid extraction

The use of a membrane as a porous barrier between the feed and the extracting solvent (i.e. SF) offers several advantages. Some of these are:

1. Efficient segregation of raffinate and extract
2. Compactness of apparatus
3. Large interfacial area
4. Constant interfacial area which is independent of the fluid velocity

The use of membranes is more successful in liquid-SF extraction than in liquid-liquid extraction since the SF can easily penetrate through the membrane into the feed phase, extract the solute and diffuse back with it to the extract side. A new supercritical fluid process called porocritical fluid extraction has been commercialized. The SF and the feed liquid flow counter-currently through a module containing a porous membrane, typically having 0.2-micron pores. Hollow fiber and spiral wound membranes are preferred since they provide high interfacial area. Fig. 7.22 shows the principle of porocritical fluid extraction.

Solid-SF extraction

The high penetrability of a SF makes it ideally suited for leaching processes. With conventional solid-liquid extraction, the solid particles need to the pulverized for satisfactory solute recovery by extraction. Particle size reduction much less important in solid-SF extraction. Such processes are usually carried out using packed beds. Fig. 7.23 shows the set-up used for extraction of caffeine from coffee beans. The extractor is first packed with coffee beans and supercritical carbon dioxide which is used as the extracting solvent is passed through this. The extract obtained consists mainly of caffeine dissolved in supercritical carbon dioxide. Caffeine and carbon dioxide are then separated by pressure reduction and the gaseous carbon dioxide thus obtained is recycled. The raffinate, i.e. caffeine free coffee beans, is suitable for producing 'decaf' coffee.

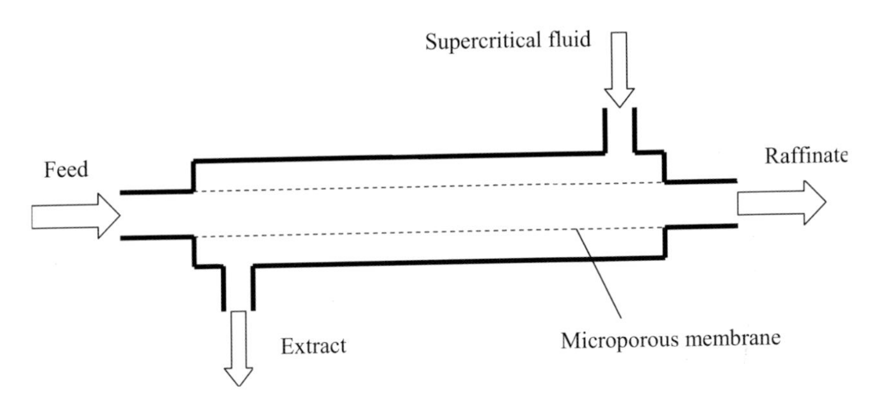

Fig. 7.22 Porocritical fluid extraction

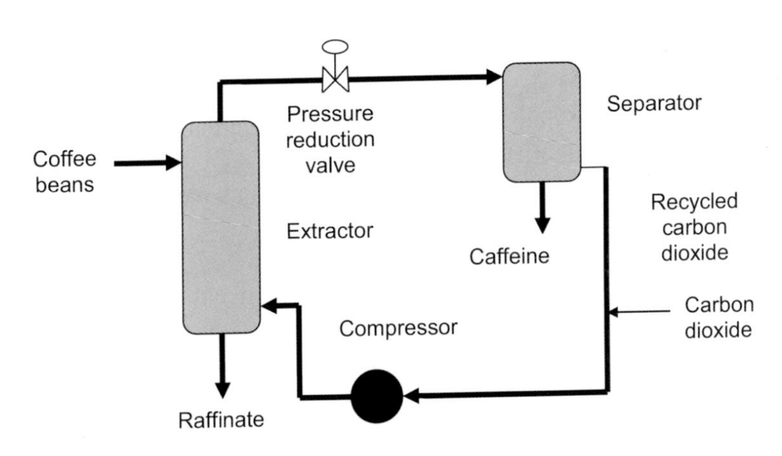

Fig. 7.23 Extraction of caffeine using supercritical carbon dioxide

Exercise problems

7.1. 10 litres of a dilute aqueous solution of a hormone (concentration = 0.1 g/l) was contacted with 1 litre of an organic solvent at 20 °C. The solute concentration in the extract thus obtained was found to be 0.7 g/l. What would have been the solute concentration in the extract had the extraction been carried out at 4°C? State assumptions required.

7.2. A solute is to be extracted from an aqueous phase to an organic phase using a continuous extractor. The aqueous feed enters the extractor at a flow rate of 400 l/h and has a solute concentration of 0.5 g/l. The equilibrium relationship is given by $C_E = 2\ C_R^2$ where C_R and C_E are the concentrations in the raffinate and extract respectively. The extraction unit consists of 2 theoretical counter-current stages. What organic solvent flow rate will be required for 60 % recovery of solute?

7.3. A differential extractor is being used to extract an amino acid from an aqueous solution (concentration = 1 g/l) to an organic solvent. The extractor is 1 m long, the feed flow rate is 0.01 m^3/s, the extracting solvent flow rate is 0.005 m^3/s, the average concentration difference across the stagnant film on the raffinate side is 0.1 g/l, and the raffinate concentration is 0.2 g/l. Calculate the mass transfer coefficient on the raffinate side if the interfacial area per unit height of the extractor is 100 m^2/m. State assumptions made.

7.4. Penicillin is to be extracted in the batch mode from 300 litres of filtered fermentation media (antibiotic concentration = 1 g/l) to 40 litres of MIBK. The equilibrium relationship is given by $C_E = ((25\ C_R) / (1+ C_R))$, where C_R and C_E are in mg/ml. Determine the percentage recovery of the antibiotic.

References

P.A. Belter, E.L. Cussler, W.-S. Hu, Bioseparations: Downstream Processing for Biotechnology, John Wiley and Sons, New York (1988).

J.M. Coulson, J.F. Richardson, J.R. Backhurst, J.H. Harker, Coulson and Richardson's Chemical Engineering vol 2, 4th edition, Butterworth-Heinemann, Oxford (1991).

C. J. Geankoplis, Transport Processes and Separation Process Principles, 4th edition, Prentice Hall, Upper Saddle River (2003).

W.L. McCabe, J.C. Smith, P. Harriott, Unit Operations of Chemical Engineering, 7th edition, McGraw Hill, New York (2005).

J.D. Seader, E.J. Henley, Separation Process Principles, 2nd edition, John Wiley and Sons, New York (2006).

M.L Schuler, F. Kargi, Bioprocess Engineering: Basic Concepts, 2nd edition, Prentice Hall, Upper Saddle River (2002).

Chapter 8

Adsorption

8.1. Introduction

Adsorption refers to the binding of molecules on the surface of solid material, i.e. an adsorption process involves the transfer of dissolved solutes from a liquid phase to the surface of an added solid phase. Adsorption is a surface phenomenon and should not be confused with absorption, which refers to the penetration of substances into the porous structure within solid material. Adsorption can be used to separate a molecule from a complex mixture of molecules, or simply to separate a solute from its solvent (see Fig. 8.1). This is achieved by contacting the solution with the solid material which is also called the adsorbent. The molecule that binds on the adsorbent is referred to as the adsorbate. The solute binding usually takes place on specific locations on the surface of the adsorbent. These are referred to as binding sites or ligands. The term ligand refers specifically to a molecular entity that has been grafted onto the surface of an inert solid material to prepare the adsorbent. Affinity binding is a selective form of adsorption where highly specific interactions between molecules and their ligands are utilized. In an adsorption based separation process, the bound material subsequently needs to be released, i.e. the binding process should be reversible in nature. The release of adsorbed material from an adsorbent is called desorption which is favoured at conditions opposite to those that favour adsorption.

Adsorption is a selective process and this is influenced by the following:

1. Molecular weight, size and shape of solute
2. Shape of the binding site or ligand
3. Polarity of the molecules and the adsorbent
4. Electrostatic charge on the molecule and on the adsorbent

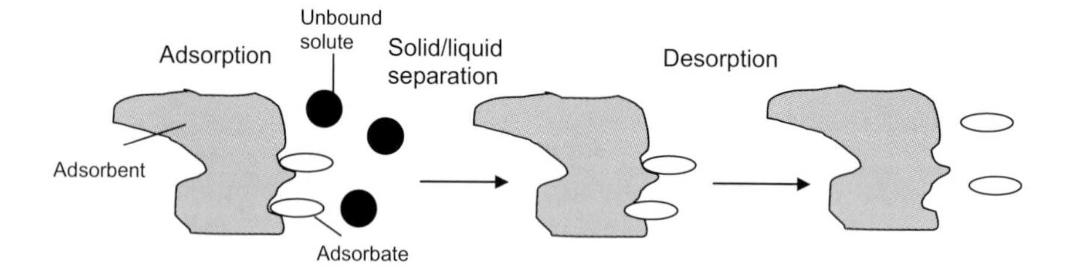

Fig. 8.1 Adsorptive separation

Some of the advantages of adsorption are:

1. High selectivity of separation (e.g. as in affinity adsorption)
2. Ability to recover solutes from dilute solutions

The disadvantages of adsorption are:

1. It is a surface phenomenon i.e. solute binding is restricted to the surface only
2. Adsorption is cyclic in nature and hence separation is carried out in batch mode only
3. Adsorption may result in loss of biological activity of the solute
4. Adsorbents can get fouled by irreversible binding of biological material

Some of the common applications of adsorption are listed below:

1. Protein purification
2. Nucleic acid purification
3. Antibiotic purification
4. Biomedical analysis
5. Pulse or elution chromatography

8.2. Adsorbents

Most adsorption processes utilize particulate adsorbents. These can be made from natural or synthetic material. Most adsorbents have amorphous or microcrystalline structure. These physical forms result in very high specific surface areas (i.e. surface area per unit amount of adsorbent). Commonly used adsorbents in bioseparation processes include cellulose based adsorbents, silica gel based adsorbents, synthetic resins, agarose based adsorbents and cross-linked dextran based adsorbents. Particulate adsorbents can be used in the form of a suspension or in the form of a packed bed. In recent years fluidized and expanded beds are increasingly being used to overcome some of the limitations associated with packed beds. More recently, membranes and monoliths are also being utilized as adsorbents.

8.3. Separation mechanisms

The physical binding of a molecule onto an adsorbent takes place due to one or more of the non-covalent interactions listed below:

1. van der Waals forces
2. Electrostatic interactions
3. Hydrophobic interactions
4. Hydrogen bonding
5. Partitioning

Depending on the type/s of interaction/s involved, separation mechanisms can be classified into several categories:

Ion exchange

Ion exchange is based on electrostatic interactions between the molecule and the adsorbent (see Fig. 8.2). A cation exchange adsorbent it itself negatively charges and can therefore bind positively charged molecules. On the other hand an anion exchange adsorbent is positively charged and can bind negatively charged molecules.

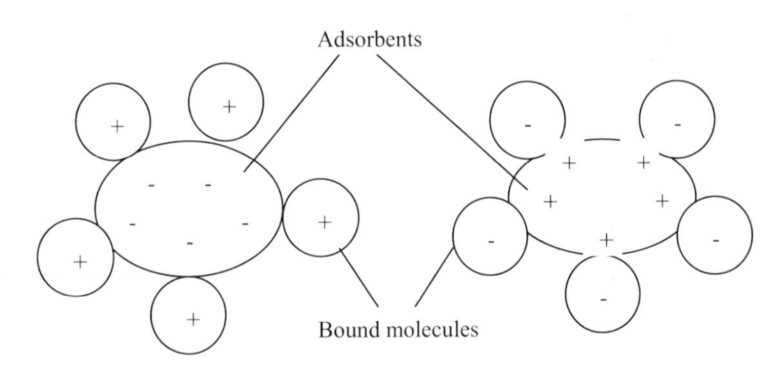

Fig. 8.2 Ion exchange adsorption

Ion exchange adsorbents are prepared by attaching charged groups onto insoluble support material. Examples of support material include cellulose, cellulose derivatives, agarose, acrylic resins and cross-linked

Agarose is the most widely used support material for preparing hydrophobic interaction adsorbents. The hydrophobic patches on the surface on the adsorbent are created by grafting alkyl and aromatic hydrocarbon groups on the agarose. Commonly used groups include butyl, octyl and phenyl.

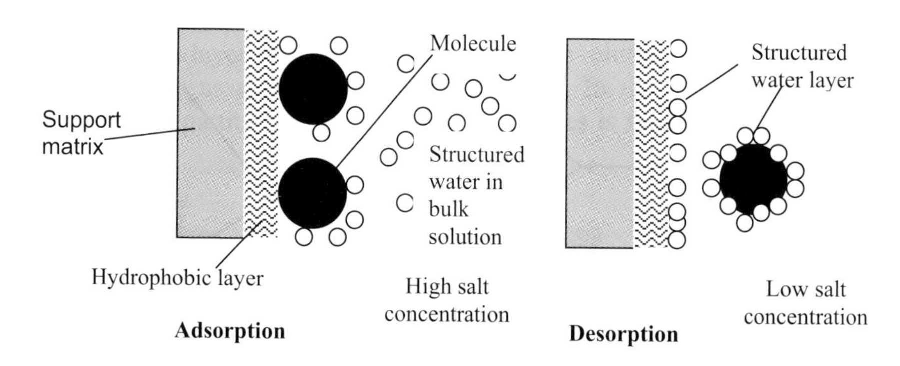

Fig. 8.9 Hydrophobic interaction based adsorption

Hydrophobic interaction based adsorption of proteins take place at very high anti-chaotropic salt concentrations, typically 1 M or higher. The salt concentration required for the adsorption of a particular protein depends on the type of salt used. When sodium sulphate is used, a lower salt concentration is required than when using ammonium sulphate or sodium chloride. Desorption of bound protein from hydrophobic interaction adsorbent can be achieved simply by lowering the salt concentration. Fig. 8.10 summarizes the separation of recombinant epidermal growth factor (rEGF) by hydrophobic interaction based adsorption.

8.4. Adsorption isotherms

The analysis of an adsorption process is based on identifying an equilibrium relationship between bound and free solute and performing a solute material balance. The equilibrium relationship between the solute concentration in the liquid phase and that on the adsorbent's surface at a given condition is called an isotherm. The solute concentration is usually expressed as the mass or number of moles of solute per unit mass or volume of solvent (or adsorbent). From the point of view of bioseparations, three types of isotherms (i.e. linear, Freundlich and Langmuir) are important (see Fig. 8.11).

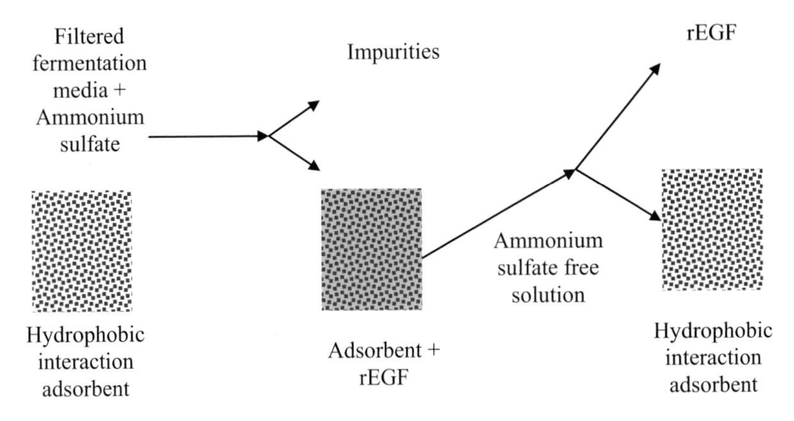

Fig. 8.10 Hydrophobic interaction separation of rEGF

Linear isotherm

Most isotherms are linear when the solute concentration is very low. Therefore when dealing with adsorption of solutes from very dilute solutions it is often convenient to use a linear expression of the form:

$$C_B = KC_U \quad (8.1)$$

Where

C_B = solute bound per unit amount of adsorbent (e.g. kg/m^3)
C_U = unbound solute concentration (in solution) (e.g. kg/m^3)
K = linear equilibrium constant (e.g. (-))

K is analogous to the partition coefficient term used in extraction. Its unit depends on the units of the two concentration terms used in equation (8.1).

Freundlich isotherm

This is an empirical correlation of the form shown below:

$$C_B = K_F C_U^{\,n} \quad (8.2)$$

Where

K_F = Freundlich adsorption constant

In equation (8.2) n is less than 1. The Freundlich isotherm does not predict any saturation of the binding surface by the solute. The adsorption of antibiotics, steroids and hormones onto commonly used adsorbents follow this type of isotherm. Reverse phase and hydrophobic interaction type adsorption generally follow Freundlich type isotherm. The constants K and n must be determined experimentally from C_B-C_U data (see Fig 8.12).

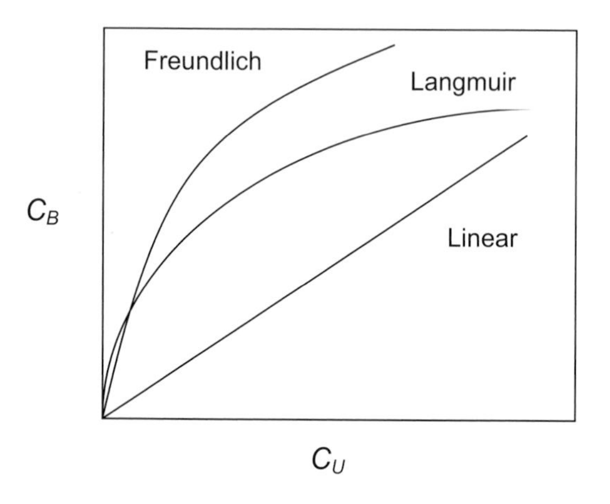

Fig. 8.11 Adsorption isotherms

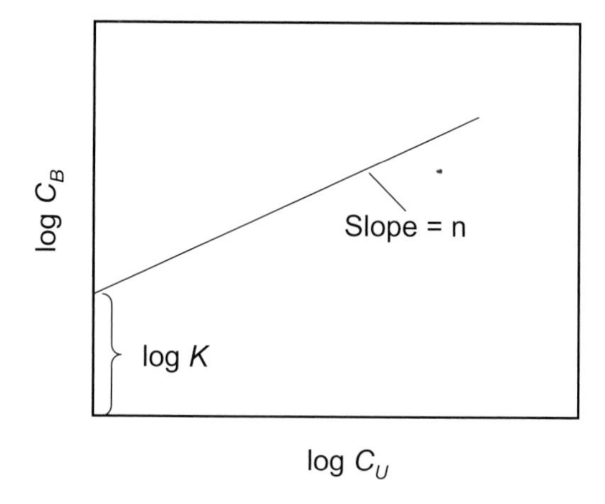

Fig. 8.12 Determination of Freundlich isotherm constants

Langmuir isotherm

This isotherm is of the form shown below:

$$C_B = \frac{C_{B\max} C_U}{K_L + C_U} \tag{8.3}$$

Where

C_{Bmax} = saturation constant (e.g. kg/m^3)

K_L = affinity constant for Langmuir isotherm (e.g. kg/m^3)

The Langmuir isotherm is applicable when there is a strong specific interaction between the solute and the adsorbent. Ion exchange and affinity type adsorptions generally follow Langmuir isotherm. This isotherm predicts the saturation of the adsorption sites by solute molecules, indicating monolayer formation. $C_{B\max}$ and K_L are constants, with K_L having the same units as C_U, and $C_{B\max}$ having the same units as C_B. K_L and $C_{B\max}$ are experimentally determined as shown in Fig. 8.13.

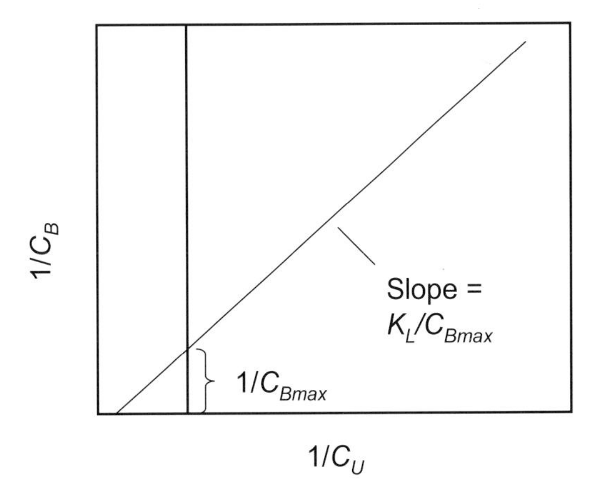

Fig. 8.13 Determination of Langmuir isotherm constants

8.5. Diffusional limitations in adsorption processes

Adsorption processes could sometimes be limited by the rate of solute transport from the feed solution to the surface of the adsorbent. The size and geometry of the adsorbent material frequently plays a significant role in adsorption kinetics. Two solute transport-limiting mechanisms are commonly observed in adsorption processes (see Fig. 8.14):

Lysozyme concentration in solution (g/l)	Adsorbent #1 Bound lysozyme concentration (g/l)	Adsorbent #2 Bound lysozyme concentration (g/l)
0.2	15	10
0.4	18	15
0.6	20	20
0.8	22	25
1.0	23	28

Comment on the nature of the adsorption isotherms. Which adsorbent will be better for the above mentioned separation?

Solution

To test for linear adsorption isotherm C_B is plotted against C_U. For Freundlich adsorption isotherm, ln (C_B) is plotted against ln (C_U). For Langmuir isotherm, ($1/C_B$) is plotted against ($1/C_U$). Based on the regression coefficients, it may be found out that both adsorptions follow Freundlich isotherm.

For adsorbent #1, the isotherm is:

$$C_B = 23.1 C_U^{0.269} \tag{8.a}$$

For adsorbent #2, the isotherm is:

$$C_B = 28.14 C_U^{0.653} \tag{8.b}$$

Lysozyme makes up 5% of the feed proteins. Therefore its concentration in the feed solution is 5% of 10 g/l, i.e. 0.5 g/l.

S = 1000 l

A = 100 l

From a lysozyme material balance:

$$C_B = \frac{1000}{100}(0.5 - C_U) \tag{8.c}$$

Equations (a) and (b) are the equilibrium lines for adsorbent #1 and # 2 respectively. Equation (c) is the common operating line. From the point of intersection of equations (a) and (c) the bound and free lysozyme concentrations with adsorbent #1 can be determined:

C_B = 4.99 g/l, C_U = 0.003 g/l

From the point of intersection of equations (b) and (c) the bound and free lysozyme concentrations with adsorbent #2 can be determined:

C_B = 4.4 g/l, C_U = 0.058 g/l

Based on this information, we can infer that adsorbent #1 is better.

Example 8.2

We are planning to recover an antibiotic from 10 litres of feed solution by adsorption using activated carbon. The concentration of the antibiotic in the feed is 1.1×10^{-6} g per g of water. Adsorption data obtained is shown below:

$C_U \times 10^6$ (g solute / g water)	$C_B \times 10^3$ (g solute / g carbon)
0.1	1.3
0.3	1.7
0.6	2.3
0.9	2.4
1.2	2.6

Which isotherm fits the data best? How much adsorbent is required for 95 % recovery of the antibiotic if this particular type of isotherm is assumed?

Solution

Making plots similar to those in the problem 8.1, it may be found out that the above adsorption follows Freundlich isotherm. The isotherm is:

$$C_B = 2.51 C_U^{0.288} \tag{8.d}$$

Antibiotic concentration in the feed = 1.1×10^{-6} g/g

For 95% recovery $C_U = 5.5 \times 10^{-8}$ g/g

Therefore (from the isotherm):

$C_B = 0.0204$ g/g

$S = 10$ l = 10,000 g

From equation (8.6), we can write:

$$A = \frac{10{,}000}{0.0204}\left(1.1\times10^{-6} - 5.5\times10^{-8}\right) \text{ g} = 0.512 \text{ g}$$

Example 8.3

A hormone is being recovered from 10 litres of a biological fluid by affinity adsorption. The adsorption follows linear isotherm and the concentration of the hormone in the fluid is 0.01 g/l. 80% of the hormone could be adsorbed in the batch mode by 10 ml of affinity adsorbent. How much hormone would be adsorbed if 50 ml of adsorbent were used?

Solution

Total amount of hormone in the feed = 10×0.01 g = 0.1 g

The amount bound to the adsorbent = 0.1×0.8 g = 0.08 g

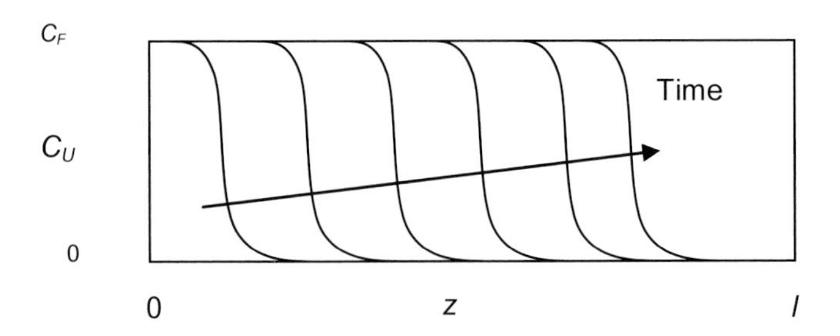

Fig. 8.19 Solute concentration profile within pack bed

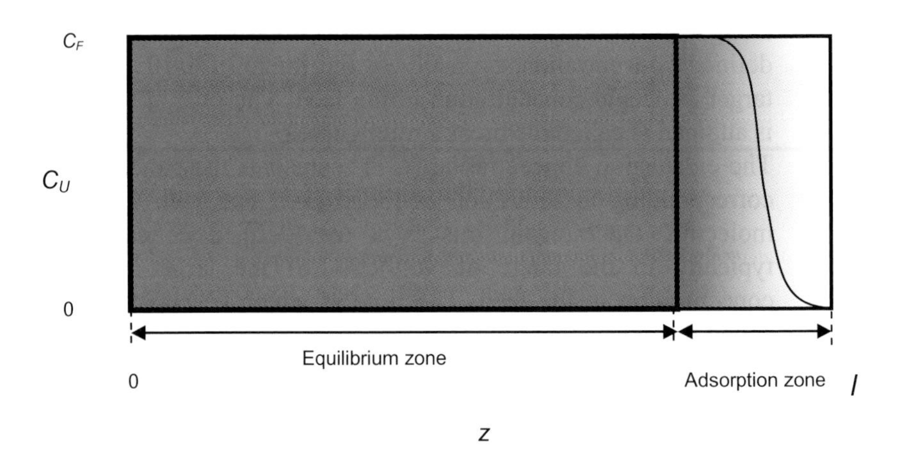

Fig. 8.20 Adsorption and equilibrium zones

The adsorption process within a packed bed is described by the partial differential equation shown below:

$$u\frac{\partial C_U}{\partial z} + \varepsilon\frac{\partial C_U}{\partial t} + (1-\varepsilon)\frac{\partial C_B}{\partial t} = E\frac{\partial^2 C_U}{\partial z^2} \tag{8.7}$$

Where

u = fluid velocity within the packed bed (m/s)

ε = voidage fraction (-)

E = axial dispersion coefficient (m^2/s)

The z axis is along the length of the bed. The length of the packed bed is equal to l. Equation (8.7) can be solved in different ways depending on the specific system under consideration. In bioseparation processes involving macromolecules axial dispersion can frequently be ignored. For an adsorption process where E is negligible and the isotherm is linear (i.e. $C_B = K\ C_U$), the breakthrough curve can be obtained using the following boundary conditions:

At $t = 0$ and $z = 0$, $C_U = C_F$
$t = t''$ and $z = l$, $C_U = C_F$
$t = t'$ and $z = l$, $C_U = B\ C_F$ where $B = (C'/C_F)$

The solution of equation (8.7) using the above boundary conditions gives the equation of the breakthrough curve:

$$C_U = C_F \exp\left[\frac{\ln B}{t''-t'}(t'-t)\right] \quad \text{for } t = 0 \text{ to } t'' \tag{8.8}$$

$$C_U = C_F \quad \text{for } t > t'' \tag{8.8a}$$

The exhaustion time can be approximated by:

$$t'' = \frac{l}{u\varepsilon}(K + \varepsilon - K\varepsilon) \tag{8.9}$$

Where axial dispersion is significant, the equation of the breakthrough curve is shown below:

$$\frac{C_U}{C_F} = 0.5\left\{1 + erf\left[\left(\frac{ul}{4E}\right)^{0.5} \frac{(t-t'')}{(tt'')^{0.5}}\right]\right\} \tag{8.10}$$

A simpler analysis of packed bed adsorption is possible based on the assumption that velocity of the target molecule concentration profile through the packed bed is constant. We need to further assume that the shape of the concentration profile within the packed bed does not change with time. Therefore we can write:

$$u = \frac{l}{t'} \tag{8.11}$$

The time taken for the concentration profile to move the hypothetical distance from breakthrough to exhaustion is $(t''-t')$ and the distance covered during this time is the length of the adsorption zone (l_{Ad}). Therefore:

$$l_{Ad} = \frac{l(t''-t')}{t'} \tag{8.12}$$

The length of the equilibrium zone is given by:

$$l_{Eq} = l\left(1 - \frac{t''-t'}{t'}\right) \tag{8.13}$$

The concentration of adsorbed target molecule within the equilibrium zone is assumed to be in equilibrium with the concentration in the feed. The average concentration of adsorbed target molecule within the adsorption zone is approximated to be half of that in the equilibrium zone. We also assume that the adsorption follows a linear isotherm which is a valid assumption for adsorption from dilute solutions. Based on this, the fraction of the binding capacity of the packed bed utilized at t' is given by:

$$\theta = 1 - \frac{t'' - t'}{2t'} \tag{8.14}$$

Equation (8.14) implies that the sharper the breakthrough curve (i.e. the smaller the difference between t'' and t') the better is the packed bed-capacity utilization. The nature of the breakthrough curve depends mainly on the aspect ratio of the bed, the quality of packing, the particle size distribution of the adsorbent, the packed bed inlet flow distribution and dispersion within the packed bed. For a given packed bed, the breakthrough curve depends on the feed flow rate. For molecules having low diffusivities, lower flow rates result in sharper breakthroughs. This is however not the case with highly diffusible molecules.

Example 8.4

A protein is being adsorbed in a 1.5 m long packed bed containing DEAE-cellulose particles. The breakthrough and exhaustion times are 8 hours and 10 hours respectively. What are the lengths of the equilibrium zone and adsorption zone at the point of breakthrough? What fraction of the bed capacity is utilized at breakthrough?

Solution

From equation (8.12):

$$l_{Ad} = \frac{1.5 \times (10 - 8)}{8}\ \text{m} = 0.375\ \text{m}$$

From equation (8.13):

$$l_{Eq} = 1.5\left(1 - \frac{10 - 8}{8}\right)\ \text{m} = 1.125\ \text{m}$$

From equation (8.14):

$$\theta = 1 - \frac{10 - 8}{2 \times 8} = 0.875$$

Example 8.5

An antibody is being adsorbed by an affinity adsorbent contained in a 5 cm long packed bed. The breakthrough time was determined to be 30 minutes and it is known that 80% of the binding capacity was utilized at this point. Predict the exhaustion time.

Solution

At breakthrough, 80% of the bed capacity was utilized. Therefore:

$\theta = 0.8$

From equation (8.14):

$$0.8 = 1 - \frac{t_E - 30}{2 \times 30}$$

Therefore:

$t_E = 42\,\text{min}$

8.8. Other types of adsorption devices

Some of the major problems associated with packed beds are:

1. High pressure drop across the packed bed i.e. high pumping cost
2. Inability to handle feed with particulate material
3. Need to use low flow rates
4. Radial dispersion

Some of these problems could be overcome using fluidized bed or expanded bed adsorbers (Fig. 8.21). In fluidized or expanded bed systems, the adsorbent is loosely packed within a column with some space on the top. The feed is pumped into the column from the bottom at a velocity sufficient to fluidize the adsorbent bed. Membrane adsorbers which use stacks of adsorptive membranes can also solve some of these problems (see Fig. 8.22). Membrane adsorbers are discussed in details in the chapter on *membrane based bioseparation*.

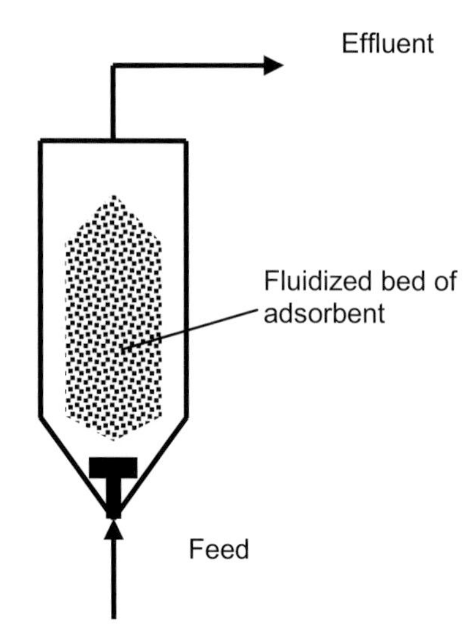

Fig. 8.21 Fluidized bed adsorber

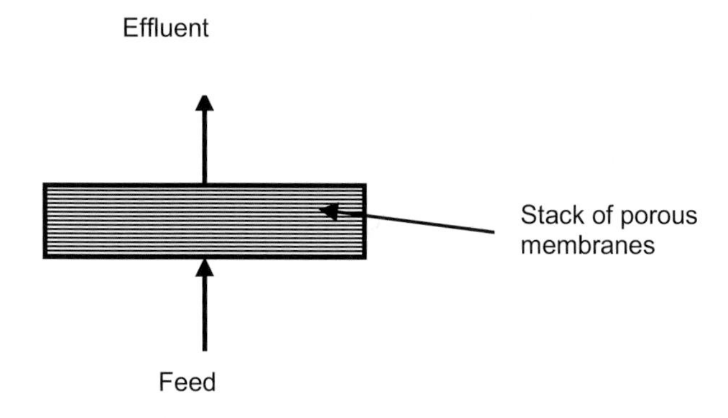

Fig. 8.22 Membrane adsorber

Exercise problems

8.1. A monoclonal antibody is being recovered from filtered cell culture media by affinity adsorption on Sepharose protein-A adsorbent contained in a packed bed (length = 5 cm, voidage fraction = 0.25). The adsorption follows linear isotherm (C_B = 15 C_U). If breakthrough of monoclonal antibody was observed after 35 minutes using a superficial feed velocity of 5 cm/min, estimate the fraction of binding capacity utilized at the point of breakthrough.

8.2. Solid spherical pellets of 2 mm radius are being used in a stirred tank to adsorb a solute from an aqueous solution. The adsorption follows a linear isotherm (C_B = 10 C_E) and the transfer of solute from bulk solution to the surface of the adsorbent is diffusionally limited. At a particular instant: the solute concentration in the bulk solution was measured to be 0.1 kg/m^3; the bound concentration was determined to be 0.5 kg/m^3; the rate of solute transfer to the adsorbent was estimated to be 0.03 kg/m^3 s; and the boundary layer was found to be approximately 10 micrometers thick. Estimate the diffusivity of the solute.

8.3. A 0.5 m long column was packed with dry silica gel beads (true density = 1264 kg/m^3) such that the bulk density of the packed bed was 670 kg/m^3. Feed containing 0.00279 kg of solute per kg of solvent enters the bed at a rate of 10 kg solvent per second per m^2 cross-sectional area. The density of the solvent is 866 kg/m^3 while the equilibrium relationship can be approximated by C_B = 3 C_U. Determine the breakthrough time if it is known that 90% of the bed capacity is utilized at the point of breakthrough.

8.4. Consider a batch adsorption process, in which 100 kg of a protein solution (0.05 kg of protein/kg solution), is contacted with 15 kg of ion-exchange adsorbent. The equilibrium relationship is given by:

$$C_B = \frac{1.5C_U}{0.1 + C_U}$$

Both C_U and C_B are expressed in mass ratio i.e. C_U is expressed in kg protein/kg solution, and C_B is expressed in kg protein/kg adsorbent. Determine the equilibrium concentrations and the fraction of protein adsorbed.

8.5. An impurity is to be removed from 20 litres of a pharmaceutical solution by contacting it with a batch of adsorbent resin. The adsorption follows a Langmuir type adsorption isotherm where K_L = 0.5 g/l and C_{Bmax} = 1.5 g/g. The concentration of the impurity in the solution is estimated to be 0.1 g/l. How much adsorbent will be needed to reduce its concentration to 0.01 g/l?

References

P.A. Belter, E.L. Cussler, W.-S. Hu, Bioseparations: Downstream Processing for Biotechnology, John Wiley and Sons, New York (1988).

M.R. Ladisch, Bioseparations Engineering: Principles, Practice and Economics, John Wiley and Sons, New York (2001).

E.L. Cussler, Diffusion: Mass Transfer in Fluid Systems, Cambridge University Press, Cambridge (1997).

C. J. Geankoplis, Transport Processes and Separation Process Principles, 4th edition, Prentice Hall, Upper Saddle River (2003).

J.M. Coulson, J.F. Richardson, J.R. Backhurst, J.H. Harker, Coulson and Richardson's Chemical Engineering vol 2, 4th edition, Butterworth-Heinemann, Oxford (1991).

J.D. Seader, E.J. Henley, Separation Process Principles, 2nd edition, John Wiley and Sons, New York (2006).

M.L Schuler, F. Kargi, Bioprocess Engineering: Basic Concepts, 2nd edition, Prentice Hall, Upper Saddle River (2002).

Chapter 9

Chromatography

9.1. Introduction

Chromatography is a solute fractionation technique which relies on the dynamic distribution of molecules to be separated between two phases: a stationary (or binding) phase and a mobile (or carrier) phase. In its simplest form, the stationary phase is particulate in nature. The particles are packed within a column in the form of a packed bed. The mobile phase is passed through the column, typically at a fixed velocity. A pulse of sample containing the molecules to be separated is injected into the column along with the mobile phase. The velocities at which these molecules move through the column depend on their respective interactions with the stationary phase. For instance, if a molecule does not interact with the stationary phase its velocity is almost the same as that of the mobile phase. With molecules that do interact with the stationary phase, the greater the extent of interaction, the slower is the velocity. This mode of chromatographic separation is also called pulse chromatography to distinguish it from step chromatography which is operated differently.

Chromatography is used for the separation of different substances: proteins, nucleic acids, lipids, antibiotics, hormones, sugars, etc. When used for analysis of complex mixtures, chromatography is referred to as analytical chromatography while when used to separate molecules as part of a manufacturing process, it is referred to as preparative chromatography. Some of the applications of chromatography in biotechnology are listed below.

1. Biopharmaceutical production
2. Biopharmaceutical and biomedical analysis

3. Environmental analysis
4. Foods and nutraceuticals production
5. Diagnostics
6. Process monitoring

9.2. Chromatography system

A chromatographic separation system consists of a column, mobile phase reservoir/s, pump/s, sample injector, detector/s and sometimes a fraction collector. Fig. 9.1 shows a simple chromatographic separation set-up.

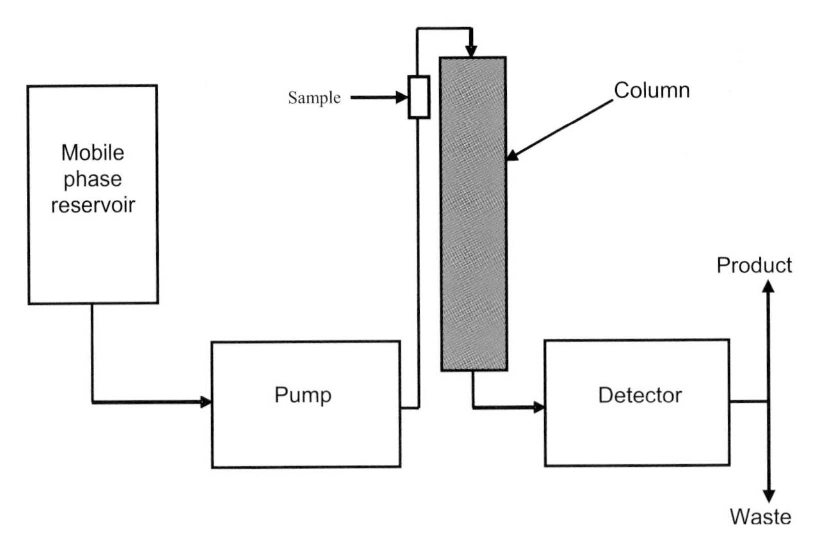

Fig. 9.1 Chromatographic system

Different types of columns are used for chromatographic separations (see Fig. 9.2). These are:

1. Packed bed column
2. Packed capillary column
3. Open tubular column
4. Membrane
5. Monolith

Packed bed columns are the most widely used type. Packed capillary and open tubular columns are mainly used for analytical chromatography of synthetic chemicals. The use of membranes in pulse chromatography is limited on account of their small bed heights. However they are increasingly being used in "bind and elute chromatography" (also called step chromatography) which is operationally similar to adsorption. Monoliths on account of their larger bed heights relative to membranes hold more promise. Their use in analytical separation is on the increase.

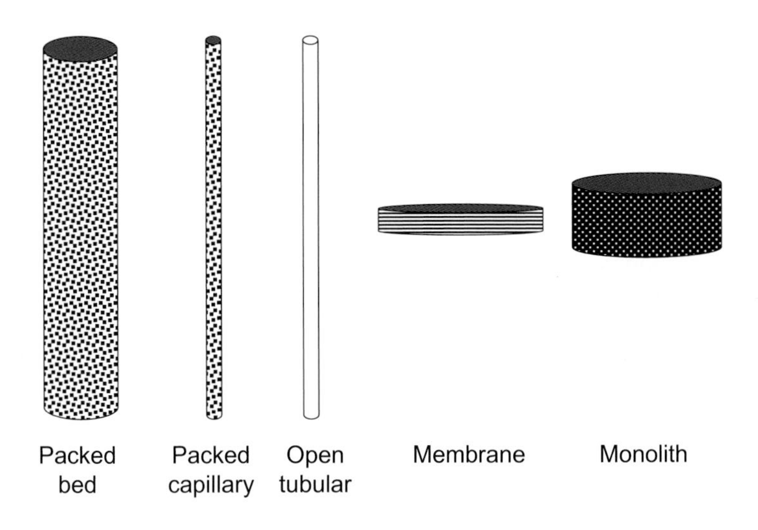

Fig. 9.2 Chromatographic columns

The different separation mechanisms used for chromatography are (also see Fig. 9.3):

1. Ion exchange
2. Reverse phase
3. Hydrophobic interaction
4. Affinity
5. Size exclusion

The first four mechanisms, i.e. ion exchange, reverse phase, hydrophobic interaction and affinity have already been discussed in the previous chapter. Ion exchange chromatography relies on the differences in electrostatic interaction between the solutes and the stationary phase as basis of separation. Reverse phase chromatography is based on the differences in the extent to which solutes partition into the non-polar stationary phase. Hydrophobic interaction chromatography separates molecules based on their differences in hydrophobicity. Affinity chromatography relies on the highly specific recognition and binding of target molecules on ligands attached to the stationary phase. Size exclusion chromatography which is also frequently referred to as gel filtration chromatography is based on the use of inert porous particles as stationary phase and these separate solutes purely on the basis of size. During their journey through the chromatographic column, smaller solute molecules find it easier to enter the pores of the chromatographic media (i.e. stationary phase) while the larger solutes are excluded form these pores. As a result of this, smaller molecules spend a longer time within the column than larger molecules and hence appear in the effluent later. The *size-exclusion limit* of a gel filtration column specifies the molecular weight range that can be resolved by that particular column. All molecules larger than the size-exclusion limit will travel at the same velocity through the column and appear in the effluent at the same time. This corresponds to the time taken by the solvent molecules to travel through the void volume within the column. With molecules smaller than the size-exclusion limit, the velocities depend on the molecular weight: larger molecules travel faster than the smaller molecules and consequently appear in the effluent earlier.

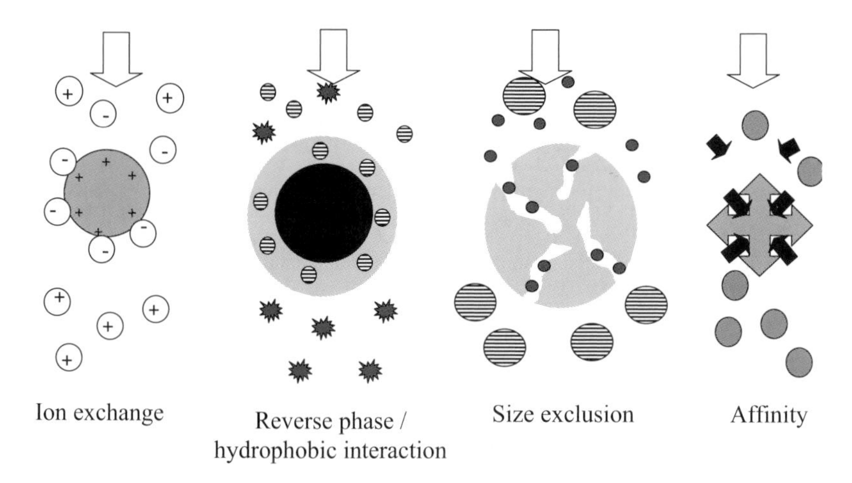

Fig. 9.3 Chromatographic separation mechanisms

The mobile phase is continuously pumped through the column during chromatographic separation. The type of pump used depends on the pressure drop across the column. Chromatographic processes are classified into the following categories depending on the pressure requirement:

1. High pressure chromatography, typically greater than 1MPa
2. Medium pressure chromatography, typically in the range of 0.1 to 1 MPa
3. Low pressure chromatography, typically less than 0.1 MPa

High pressure chromatography is frequently referred to as High Performance Liquid Chromatography or HPLC. Fig. 9.4 shows the picture of an HPLC system. Special pumps which can deliver constant mobile phase flow rates at high pressures are needed in such equipment. Plunger pumps are commonly used in HPLC systems. The pumping requirements are less demanding in medium pressure chromatography. Plunger pumps, diaphragm pumps and peristaltic pumps are frequently used. Peristaltic pumps are particularly suitable for handling sterile substances in a contamination free manner. Low pressure chromatography is usually carried out using peristaltic pumps. Where the pressure requirement is low, gravity flow can also be used.

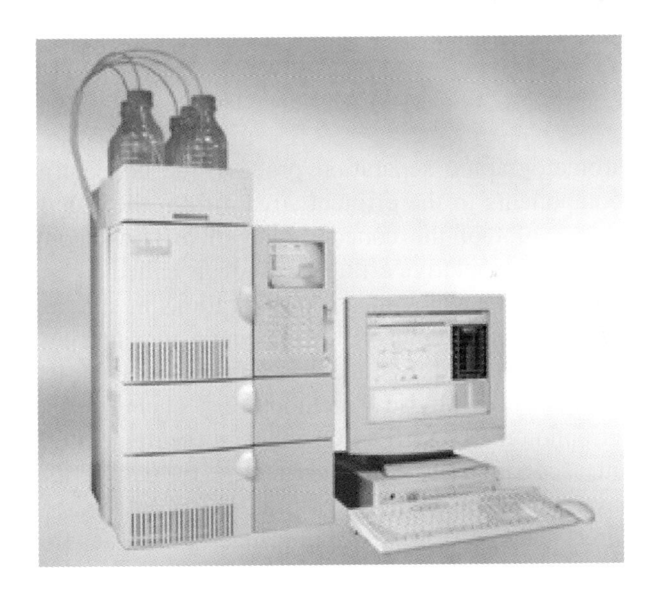

Fig. 9.4 HPLC system (Photo courtesy of Shimadzu Corporation)

The sample containing molecules to be separated is injected into the flowing mobile phase just before it enters the column. A typical sample injector consists of a motorized or manual valve and a sample loop (see Fig. 9.5). When a sample is not being injected, the mobile phase bypasses the loop as shown in the figure. When the injector is in this mode, the sample can be loaded into the sample loop. When the sample is to required to be injected, the loop is brought on line, i.e. the flow of mobile phase is directed through it. This results in the sample being carried along with the mobile phase into the column.

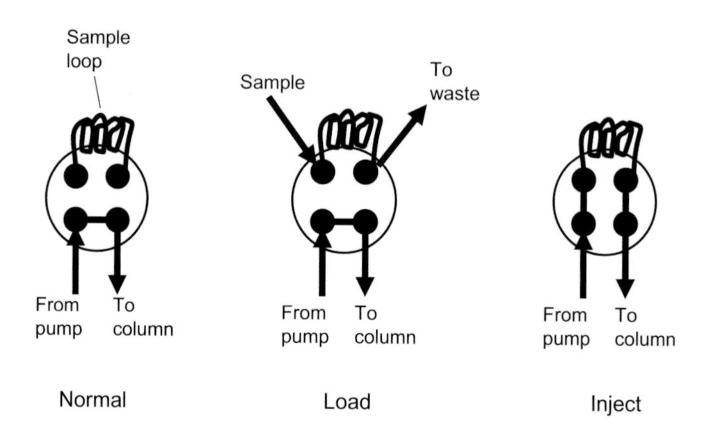

Fig. 9.5 Sample injector

In a chromatographic separation process, the concentration of the individual components in the effluent stream from the column needs to be monitored. A plot of the concentration of different components as function of time or cumulative effluent volume is called a chromatogram. There are two ways by which these concentrations in can be monitored. One way is to collect samples either manually or using an automated sample collector followed by analysis of individual fractions using appropriate chemical or physical methods (e.g. chemical analysis, UV-visible spectrophotometry, refractive index measurement, fluorescence measurement, conductivity measurement, turbidity measurement, immunoassays, etc.). This is referred to as offline analysis. The second and more direct way is to use an appropriate online (i.e. flow-through) detector for measuring a physical property, which is related to concentration. Properties that could be measured online include UV/visible absorbance, fluorescence, refractive index, conductivity and light scattering. Using appropriate calibrations, the concentration can then easily be determined. In many chromatography systems, the pH of the effluent stream and the system pressure (or sometimes pressure drop across the column) are also monitored. The monitoring of pH is important in ion exchange chromatography while keeping track of the pressure drop across the column is important from a safety point of view. The pressure drop also yields vital information about the health of the column, i.e. whether it is getting clogged up, and if so, to what extent.

The principle of chromatographic separation of a binary solute mixture is shown in Fig. 9.6. The introduction of the sample into the column is referred to as sample injection. In chromatographic separation, the instant of sample injection is referred to as zero time or zero effluent volume. Depending on the interaction of the components of the binary mixture with the stationary phase, these components move at different velocities through the column. On account of this, these components are segregated into moving bands which appear in the effluent stream as separate peaks at different times. If the concentration of the separates solutes in the effluent is plotted against time/cumulative effluent volume, a chromatogram is obtained. The time at which the maximum concentration of a component in the effluent stream is reached is referred to as its retention time.

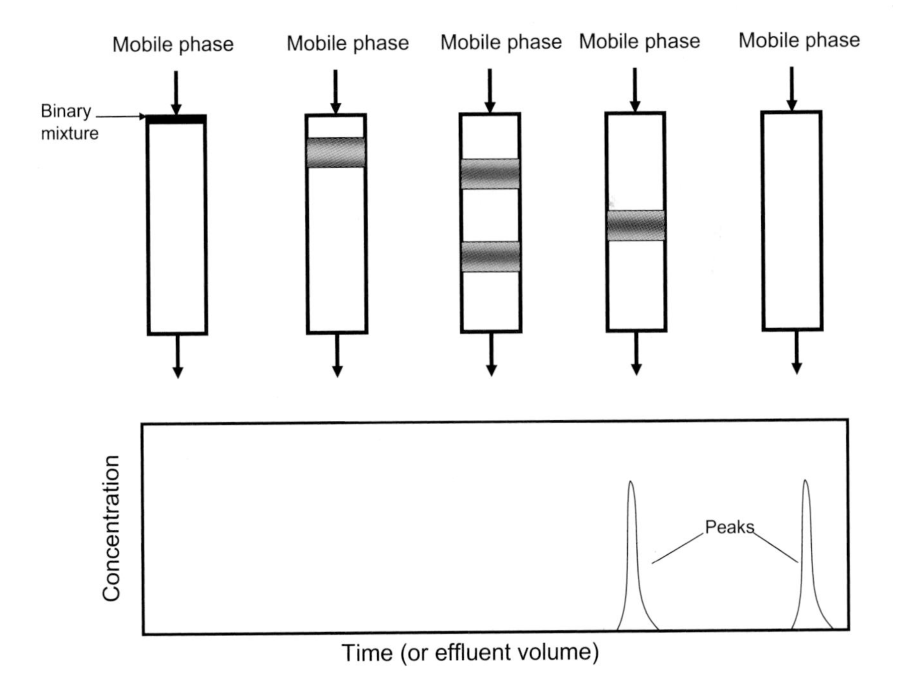

Fig. 9.6 Chromatographic separation of binary mixture

9.3. Theory of chromatography

The extent to which a molecule interacts with a stationary phase is quantified in terms of the capacity factor (k'):

$$k' = \frac{n_S}{n_M} \tag{9.1}$$

Where

n_S = moles of solute bound to the stationary phase (kg-moles)
n_M = moles of solute present in the mobile phase (kg-moles)

Therefore:

$$k' = \frac{V_S c_S}{V_M c_M} = K \frac{V_S}{V_M} \tag{9.2}$$

Where

V_S = volume of stationary phase (m^3)
V_M = volume of mobile phase (m^3)
c_S = concentration of solute in stationary phase (kg-moles/m^3)
c_M = concentration solute in mobile phase (kg-moles/m^3)
K = distribution coefficient (-)

From equations (9.1) and (9.2):

$$k' = K\left(\frac{1-\varepsilon}{\varepsilon}\right) \tag{9.3}$$

Where

ε = voidage fraction of column (-)

For a solute that does not interact with the stationary phase, the value of capacity factor is zero. For solutes that do interact with the stationary phase k' should ideally be between 1 and 10. Solutes with higher capacity factors take very long to appear in the effluent and such solutes are separated using binary or gradient chromatography. The solutes being separated by chromatography have different retention times (t_R) within the column. The retention time is the time at which the concentration of that solute reaches its maximum value in the effluent, i.e. time corresponding to the peak concentration. The retention time has two components: the mobile phase retention time (t_M) within a column which is based purely on hydrodynamic considerations, and the adjusted retention time (t_R') which is due to solute-stationary phase interaction:

$$t_R = t_M + t_R' \tag{9.4}$$

Where

t_M = mobile phase retention time (s)
t_R' = adjusted solute retention time (s)

It has been shown that:

$$t_R' = k' t_M \tag{9.5}$$

From equations (9.3), (9.4) and (9.5):

$$t_R = t_M \left(1 + \frac{1-\varepsilon}{\varepsilon} K \right) \tag{9.6}$$

Therefore, the greater the value of K of a solute, the greater is its retention time. The mobile phase retention time (t_M) depends on its flow rate, the column volume and the voidage fraction:

$$t_M = \frac{V_C \varepsilon}{Q} \tag{9.7}$$

Where
Q = mobile phase flow rate (m^3/s)
V_C = column volume (m^3)

From equations (9.6) and (9.7), we get:

$$t_R = \frac{V_C}{Q} (\varepsilon + K - \varepsilon K) \tag{9.8}$$

The solute concentration in a pulse is rectangular, i.e. where present the concentration is of a constant value and elsewhere it is zero. However, after its journey through the column the solutes appears in the effluent and therefore in the chromatograms as a peak. This transformation of shape takes place due to a large number of factors:

1. Interaction with the stationary phase
2. Non-ideal inlet distribution
3. Radial dispersion
4. Axial dispersion
5. Golay-Taylor dispersion

This change in shape is primarily due to the interaction of the solute with the stationary phase. This interaction results in the dynamic distribution of the solute between the mobile phase and the stationary phase during its journey through the column. In Fig. 9.8 a chromatographic column has been visualized as consisting of five segments, within each of which the solutes can distribute between the mobile and stationary phases. In this example we assume that the capacity factor is 1. In Fig. 9.7, the darker shade represents the moles of solute in the mobile phase while the lighter shade represents the moles of bound solute. The solute distribution corresponding to five time increments is shown. As the pulse of the solute moves through the column, its shape changes to that of a peak as shown in the figure due to the manner in which it is distributed.

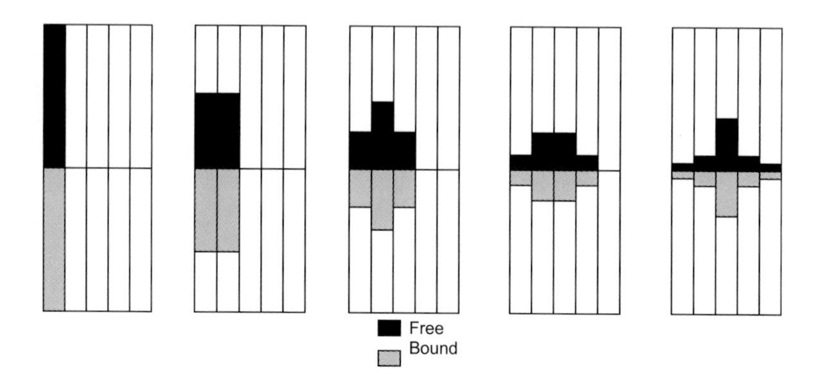

Fig. 9.7 Partitioning of a solute within a chromatographic column

Chromatographic separation primarily relies on difference in solute retention times, i.e. the solute peaks should appear separately. The quality of separation depends on how well the peaks are resolved. Fig. 9.8 shows a typical chromatogram for the separation of two solutes. Here t_{R1} and t_{R2} are the retention times of two solutes being separated while w_1 and w_2 are their characteristic peak widths. The peak width depends on the interactions between the solute and the stationary phase, the mobile phase flow rate and the solute concentration in the injected sample. The spatial separation to the two peaks (i.e. whether the peaks overlap or do not) is measured in terms of the resolution parameter:

$$R = \frac{t_{R2} - t_{R1}}{0.5(w_1 + w_2)} \tag{9.9}$$

Where

R = resolution parameter (-)

t_{R1} = retention time of solute eluted first (s)

t_{R2} = retention time of solute eluted second (s)

w_1 = peak width of solute eluted first (s)

w_2 = peak width of solute eluted second (s)

The greater the value of R, the better is the spatial separation of the peaks. An R value of less than 1 indicates that the peaks overlap, a value of 1 indicates that the peaks are just resolved, while a value greater than 1 indicates there is a gap between the separated peaks.

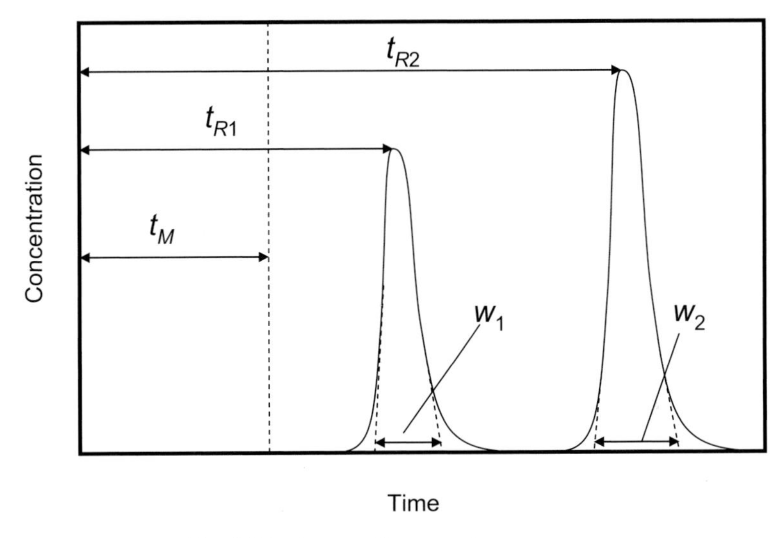

Fig. 9.8 Peak separation in a chromatogram

Chromatographic separation can also be quantified in terms of the selectivity parameter (α). This depends on the difference in interaction of the two solutes with the stationary phase and the mobile phase flow rate but is independent of the solute concentrations. The greater the value of α the better is the separation.

$$\alpha = \frac{K_2}{K_1} = \frac{k'_2}{k'_1} = \frac{t_{R2} - t_M}{t_{R1} - t_M} \tag{9.10}$$

A chromatographic column can be assumed to be made up of a large number of theoretical plates which are analogous to those in a distillation column. The greater the number of these plates in a column the better is the separation. The height equivalent of a theoretical plate (*HETP* or simply *H*) is given by:

$$H = \frac{l}{N} \tag{9.11}$$

Where

N = number of theoretical plates (-)

l = length of the column (m)

The number of theoretical plates in a chromatographic column can be determined by obtaining a chromatogram with a solute. N depends on the solute retention time and peak width as shown below:

$$N = 16\left(\frac{t_R}{w}\right)^2 \tag{9.12}$$

The theoretical plate height can be determined from the van Deemter equation:

$$H = A + \frac{B}{u} + Lu \tag{9.13}$$

Where

A = eddy diffusivity component (m)
B = axial diffusion constant (m^2/s)
L = transfer constant (s)
u = mobile phase velocity (m/s)

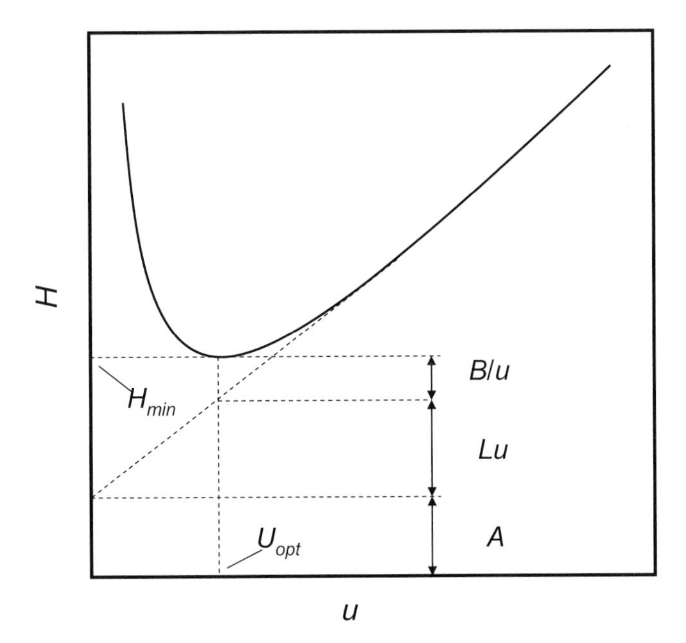

Fig. 9.9 van Deemter plot

A depends on the variability of path length followed by the mobile phase within the column. This in turn depends on the nature of column packing, particularly the particle diameter and size distribution. The smaller the particle size and particle size distribution, the smaller is the value of A. B depends on the axial diffusion of the solute in the column. Axial diffusion becomes more significant at lower mobile phase flow rates. L represents the solute transfer between mobile phase and stationary phase. Fig. 9.9 shows the van Deemter plot which is basically a plot of H versus u. At lower values of mobile phase velocity, increasing the velocity decreases the plate height. In other words, the efficiency of chromatographic separation increases with increase in velocity. In this velocity range, the second term i.e. (B/u) is dominant. The value of H decreases with increase in u to a minimum value referred to as $HETP_{min}$ or simply H_{min}. The corresponding value of u is referred to as the u_{opt} or the optimum mobile phase velocity. Further increase in velocity results in increase in the value of H. In this mobile phase velocity range, the third term i.e. (Lu) dominates. The resolution of chromatographic separation depends on the number of theoretical plates as shown below:

$$R = 0.5\sqrt{N}\left(\frac{\alpha - 1}{\alpha + 1}\right)\left(\frac{k'_{ave}}{1 + k'_{ave}}\right) \tag{9.14}$$

Where

$$k'_{ave} = 0.5(k'_1 + k'_2) \tag{9.15}$$

Example 9.1

Egg white proteins are being separated by isocratic chromatography using a 10 cm long column having 250 theoretical plates. The distribution coefficients for the proteins are given below:

Protein	Distribution coefficient
Ovalbumin	0
Conalbumin	1
Lysozyme	5

If the voidage fraction of the column is 0.45 and the mobile phase retention time is 10 minutes, predict the retention time of the three proteins. Comment on the selectivity and resolutions of separation.

Solution

The residence times of the three proteins can be obtained using equation (9.6):

coefficients for IgG and albumin are 1 and 0.1 respectively. If the albumin peak has a characteristic peak width of 0.52 minutes, predict the selectivity and resolution. When the mobile phase flow rate was increased to 20 ml/min the *HETP* was found to increase by 80%. Predict the selectivity and resolution at the higher flow rate.

Solution

The volume of the column is 39 ml (calculated using the data given)
The mobile phase retention time can be calculated using equation (9.7):

$$t_M = \frac{39 \times 0.25}{10} \text{ min} = 0.975 \text{ min}$$

The retention times can be calculated using equation (9.6):

$$t_{R,alb} = 0.975\left(1 + \frac{1-0.25}{0.25} \times 0.1\right) \text{min} = 1.27 \text{ min}$$

$$t_{R,IgG} = 0.975\left(1 + \frac{1-0.25}{0.25} \times 1\right) \text{min} = 3.9 \text{ min}$$

The number of theoretical plates in the column can be calculated from albumin retention time data using equation (9.12):

$$N = 16\left(\frac{1.27}{0.52}\right)^2 = 95$$

The peak width of IgG can be calculated using equation (9.12):

$$w_{IgG} = \frac{t_{R,IgG}}{\sqrt{\frac{N}{16}}} = \frac{3.9}{\sqrt{\frac{95}{16}}} \text{ min} = 1.6 \text{ min}$$

The resolution of separation can be calculated using equation (9.9):

$$R = \frac{3.9 - 1.27}{0.5(1.6 + 0.52)} = 2.58$$

The selectivity can be calculated using equation (9.10):

$$\alpha = \frac{1}{0.1} = 10$$

The height of a theoretical plate can be calculated using equation (9.11):

$$H = \frac{50}{95} = 0.526 \text{ cm}$$

At the higher flow rate, the height of the theoretical plate increased by 80%. Therefore:

$$H = 1.8 \times 0.526 \text{ cm} = 0.95 \text{ cm}$$

Therefore the number of theoretical plates is reduced:

$$N = \frac{50}{0.95} = 52$$

The mobile phase retention time at the higher flow rate is:

$$t_M = \frac{39 \times 0.25}{20} \text{ min} = 0.4875 \text{ min}$$

The new retention times are:

$$t_{R,alb} = 0.4875\left(1 + \frac{1-0.25}{0.25} \times 0.1\right) \text{min} = 0.633 \text{ min}$$

$$t_{R,IgG} = 0.4875\left(1 + \frac{1-0.25}{0.25} \times 1\right) \text{min} = 1.95 \text{ min}$$

The peak widths are:

$$w_{alb} = \frac{0.633}{\sqrt{\frac{52}{16}}} \text{ min} = 0.35 \text{ min}$$

$$w_{IgG} = \frac{1.95}{\sqrt{\frac{52}{16}}} \text{ min} = 1.08 \text{ min}$$

The selectivity is independent of the flow rate, i.e. is still 10. The resolution is:

$$R = \frac{1.95 - 0.633}{0.5(0.35 + 1.08)} = 1.84$$

Hence, the proteins can still be separated at the higher flow rate.

9.4. Shape and yield of a chromatographic peak

The plate theory predicts that a chromatographic elution peak would approximate a Poisson distribution. For a sufficiently large number of plates it can be further simplified to the Gaussian form shown in Fig.

9.10. This approximation yields an equation which is relatively easy to work with. The concentration profile in a peak is given by:

$$C = C_0 \exp\left(-\frac{(t - t_0)^2}{2\sigma^2}\right) \tag{9.16}$$

Where

C_0 = maximum solute concentration in peak (kg/m^3)
t_0 = time at which maximum concentration is reached (s)
σ^2 = Variance of the peak (s^2)

The characteristic peak width w is equal to 4σ. The time t_0 is the same as t_R used earlier in this chapter. Equation (9.16) is frequently written in the dimensionless form shown below (also see Fig. 9.11):

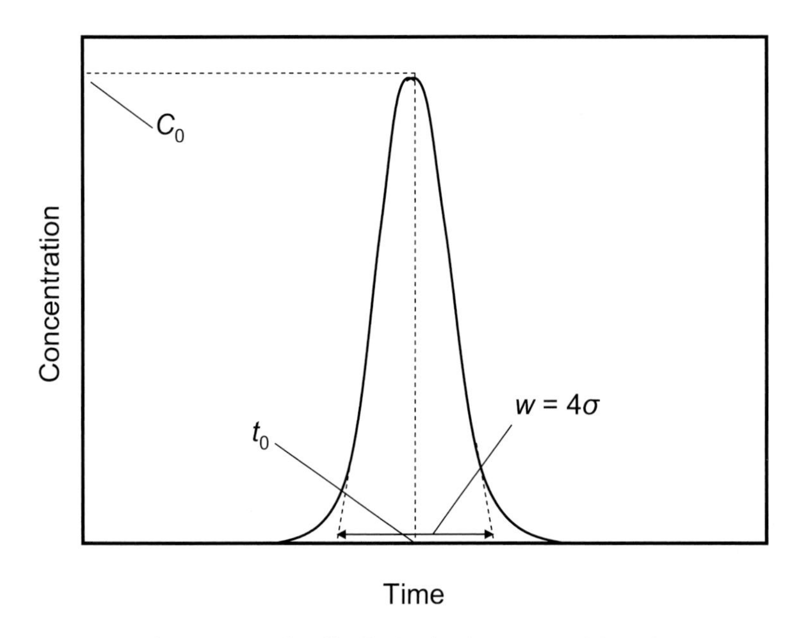

Fig. 9.10 Gaussian distribution in chromatographic peak

$$\left(\frac{C}{C_0}\right) = \exp\left(-\frac{\left(\dfrac{t}{t_0} - 1\right)^2}{\dfrac{2\sigma^2}{t_0^2}}\right) \tag{9.17}$$

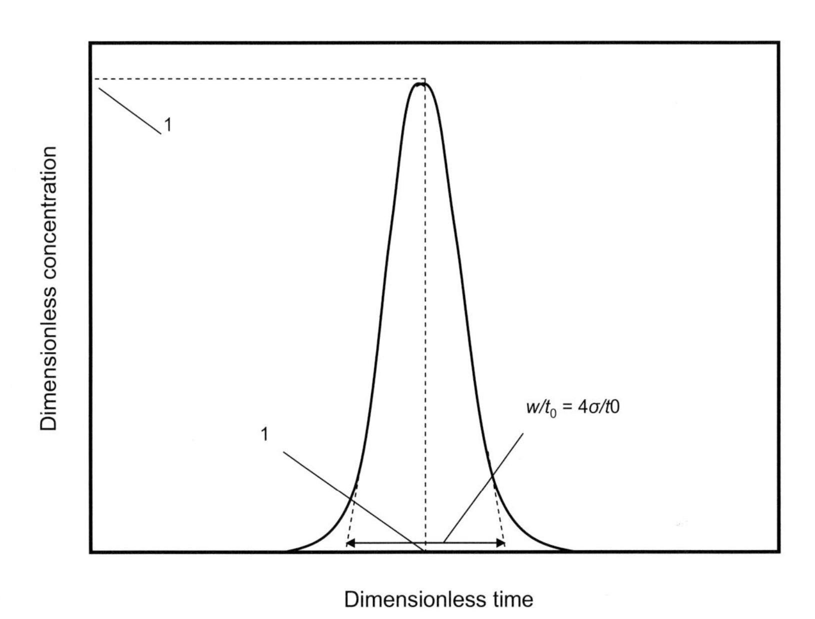

Fig. 9.11 Dimensionless plot of a chromatographic peak

The term (C/C_0) is the dimensionless concentration and the term (t/t_0) is the dimensionless time. The total amount of a solute eluted (M_{total}) in a peak is given by:

$$M_{Total} = Q\int_0^{\infty} C\,dt \tag{9.18}$$

Where

M_{total} = total amount of solute in peak (kg)

Q = mobile phase flow rate (m^3/s)

Therefore the yield of solute in the effluent collected in the time period t_1 to t_2 is:

$$\%yield = \left[\frac{\int_{t_1}^{t_2} C\,dt}{\int_0^{\infty} C\,dt}\right] \times 100 \tag{9.19}$$

The purity of a solute A in the binary separation of solutes A and B, in effluent collected in the time period t_1 to t_2 is:

$$\% purity\ A = \left[\frac{\int_{t_1}^{t_2} C_A dt}{\int_{t_1}^{t_2} C_A dt + \int_{t_1}^{t_2} C_B dt} \right] \times 100 \quad (9.20)$$

Example 9.4

Chromatograms (see figures below) were obtained for two different compounds A and B by injecting pure samples of these substances into a 30 cm long column. In both experiments, the same mobile phase flow rate was used. We would like to separate A and B at the same mobile phase flow rate from a mixture containing the same amounts of these substances as used for obtaining the chromatograms. The mobile phase residence time is 2 minutes and the voidage fraction is 0.3. Calculate:

a. The theoretical plate height of the chromatographic column
b. The selectivity
c. The resolution

If we collect the column effluent from the start to 7 minutes, calculate:

d. The purity of A in the sample
e. The percent yield of A in the sample

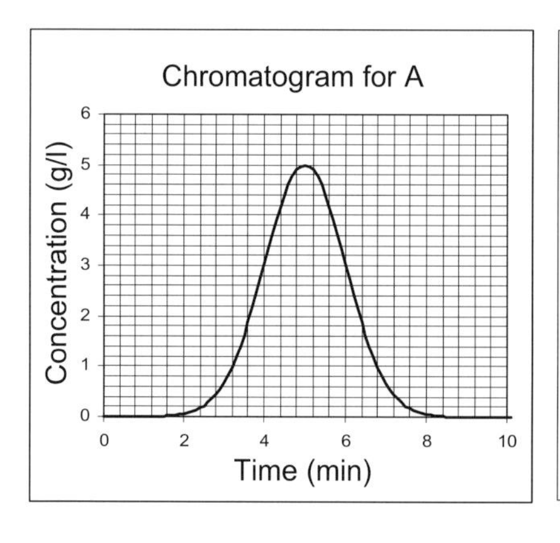

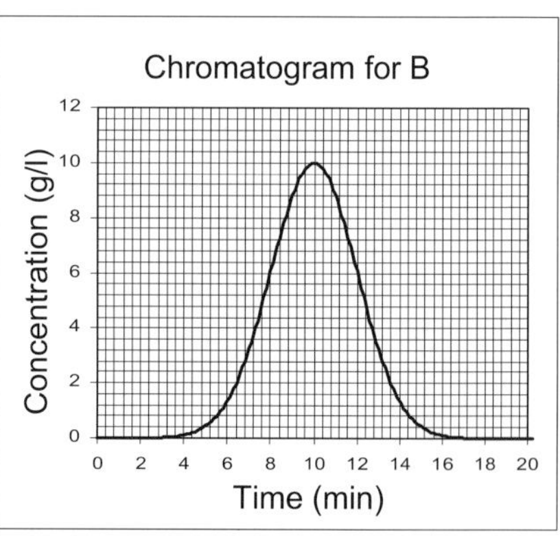

Solution

From the chromatograms:

$$w_A = 4\,\text{min}$$

$$w_B = 8\,\text{min}$$

$$t_{0,A} = 5\,\text{min}$$

$$t_{0,B} = 10\,\text{min}$$

The number of theoretical plates can be calculated using equation (9.12) based on any one of the two solutes:

$$N = 16\left(\frac{t_{0,A}}{w_A}\right)^2 = 16\left(\frac{5}{4}\right)^2 = 25$$

The plate height can be calculated using equation (9.11):

$$H = \frac{30}{25}\ \text{cm} = 1.2\ \text{cm}$$

The selectivity can be calculated using equation (9.10):

$$\alpha = \frac{10-2}{5-2} = 2.667$$

The resolution can be calculated using equation (9.9):

$$R = \frac{10-5}{0.5(4+8)} = 0.833$$

This implies that the peaks overlap.

The percentage purity of A in a sample collected from 0 to 7 minutes can be calculated using equation (9.20):

$$\%A = \frac{\int_0^7 C_A dt}{\int_0^7 C_A dt + \int_0^7 C_B dt} \times 100$$

C_A or C_B can be obtained using equation (9.16). Substituting for C using equation (9.16) in equation (9.20) we get:

$$\%A = \frac{\int_0^7 C_{0,A} \exp\left(-\left(\frac{t-5}{\sqrt{2}\sigma_A}\right)^2\right) dt}{\int_0^7 C_{0,A} \exp\left(-\left(\frac{t-5}{\sqrt{2}\sigma_A}\right)^2\right) dt + \int_0^7 C_{0,B} \exp\left(-\left(\frac{t-10}{\sqrt{2}\sigma_B}\right)^2\right) dt} \times 100$$

By solving the definite integrals in the above equation we get:

$$\%A =$$

$$\frac{\sqrt{2\pi}\sigma_A C_{0,A}\left(Erf\left(\frac{7-5}{\sqrt{2}\sigma_A}\right) + Erf\left(\frac{5}{\sqrt{2}\sigma_A}\right)\right)}{\sqrt{2\pi}\sigma_A C_{0,A}\left(Erf\left(\frac{7-5}{\sqrt{2}\sigma_A}\right) + Erf\left(\frac{5}{\sqrt{2}\sigma_A}\right)\right) + \sqrt{2\pi}\sigma_B C_{0,B}\left(Erf\left(\frac{7-10}{\sqrt{2}\sigma_B}\right) + Erf\left(\frac{10}{\sqrt{2}\sigma_B}\right)\right)} \times 100$$

$$\sigma = \frac{w}{4}$$

Therefore

$$\sigma_A = 1$$

$$\sigma_B = 2$$

From the figures:

$$C_{0,A} = 5 \text{ g/l}$$

$$C_{0,B} = 10 \text{ g/l}$$

Substituting these values in the above equation we get:

$$\% A = 78.5\%$$

From the chromatogram for solute A it may be safely assumed that all A is eluted before 10 minutes. Therefore the yield of A in sample collected from 0 to 7 minutes is given by:

$$Yield_A = \frac{\int_0^7 C_{0,A} \exp\left(-\left(\frac{t-5}{\sqrt{2}\sigma_A}\right)^2\right) dt}{\int_0^{10} C_{0,A} \exp\left(-\left(\frac{t-5}{\sqrt{2}\sigma_A}\right)^2\right) dt} \times 100$$

Solving the definite integrals as shown previously, we get:
$Yield_A = 97.7\%$

9.5. Binary chromatography

The theory and methods of chromatographic separation discussed so far is based on the use of a single mobile phase. This manner of carrying out chromatographic separation is referred to as isocratic chromatography. Size exclusion separation is carried out in the isocratic mode. For ion exchange, reverse phase, hydrophobic interaction and affinity chromatography, it is frequently convenient to use a combination of two mobile phases for increasing both selectivity and resolution. This manner of carrying out chromatographic separation using two mobile phases is referred to as binary chromatography. Fig. 9.12 shows a set-up used for binary chromatography. The general approach in binary chromatography is:

1. Use an initial mobile phase which promotes strong interactions between the stationary phase and solutes that need to be retained in the column
2. Wait till unbound or weakly interacting solutes have been removed from the column
3. Switch over to a second mobile phase in an appropriate manner to sequentially release or elute out the bound solutes

The second mobile phase in binary chromatography is selected such that in its presence the bound solutes cannot continue to interact with the stationary phase and are consequently released. The order in which the bound solutes appear in the effluent depends on:

1. The nature of the interactions between the solutes and the stationary phase
2. The properties of the second mobile phase
3. The manner in which the mobile phase is changed

Fig. 9.13 shows a typical chromatogram obtained by binary chromatography. A mixture of four solutes was injected in the pulse, A being non-interacting, B being weakly interacting, C being strongly interacting, and D being very strongly interacting. Solute A appeared in the effluent at about the same retention time as the mobile phase. This was followed by the weakly interacting component B. C and D did not appear in the effluent stream as long as the initial mobile phase was passed through the column. However, when the first mobile phase was replaced by the second mobile phase (in a linear fashion in this particular case) C and D appeared in the effluent in that order.

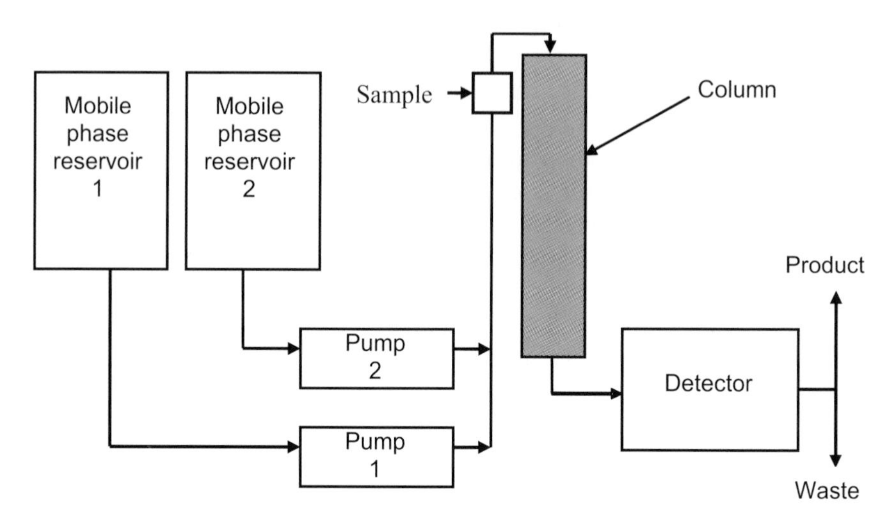

Fig. 9.12 Set-up for binary chromatography

The thumb-rules for selecting mobile phases in binary chromatography based on different separation mechanisms are listed below:

1. *Ion-exchange*
   First mobile phase: low ionic strength
   Second mobile phase: high ionic strength
2. *Affinity*
   First mobile phase: physiological conditions, sometimes higher ionic strengths
   Second mobile phase: acidic pH
3. *Reverse phase*
   First mobile phase: polar
   Second mobile phase: non-polar
4. *Hydrophobic interaction*
   First mobile phase: High concentration of anti-chaotropic salt
   Second mobile phase: Anti-chaotropic salt free

The changeover of mobile phase can be carried out in different ways (also see Fig. 9.14):

1. Step change
2. Linear gradient
3. Non-linear gradient

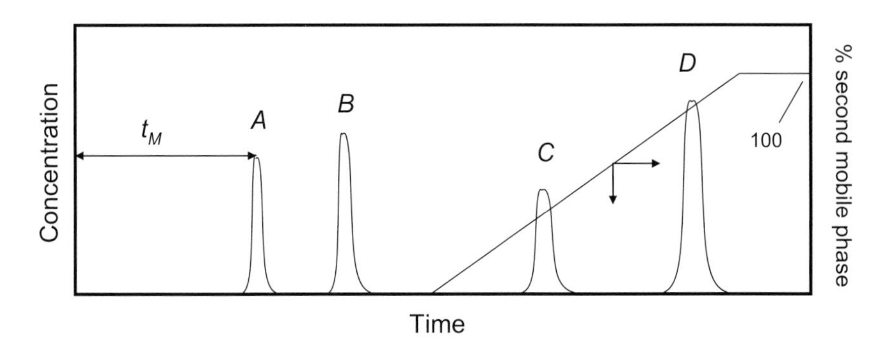

Fig. 9.13 Separation of solutes by binary chromatography

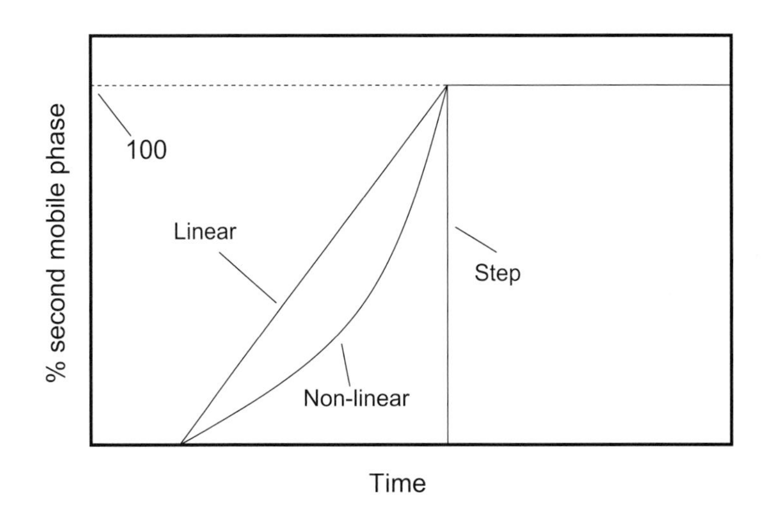

Fig. 9.14 Changeover of mobile phase in binary chromatography

9.6. Hydrodynamic chromatography

Hydrodynamic chromatography (or HDC) is similar to size exclusion chromatography (SEC) in that it also does size-based separation. However, unlike SEC which employs a packed bed, HDC uses an open tubular capillary column. Fig. 9.15 shows the principle of hydrodynamic chromatography. The mobile phase flows through the column in a streamline (or laminar) fashion and hence has a parabolic velocity distribution. If a pulse of two different solutes (one larger than the other) is introduced into the column, these will quickly align themselves along different velocity streamlines. The larger solute on account of its smaller exclusion circle will remain closer to the centerline while the smaller solute on account of its larger exclusion circle will have greater access to the streamlines closer to the wall. As a result of this, the average velocity of the smaller solute will be lower than that of the larger solute. Therefore the larger solute will appear in the column effluent as a peak earlier than the smaller one. Hydrodynamic chromatography is mainly used for separation of large macromolecules and particles.

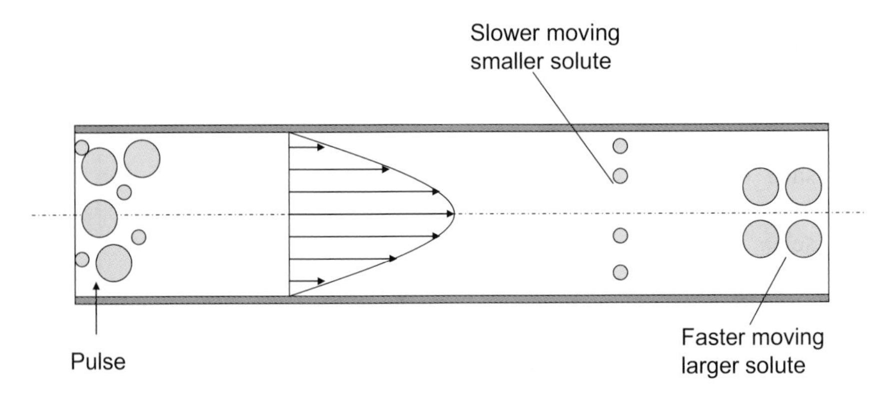

Fig. 9.15 Hydrodynamic chromatography

Exercise problems

9.1. A chromatographic column having a volume of 2.5 liters is found to have a mobile phase retention time of 2 minutes at a flow rate of 0.5 l/min. A pulse of a pharmaceutical protein was injected into the column along with the mobile phase and using appropriate techniques the concentration of bound protein and free protein at a particular slice of the column at a particular instant were found to be 0.73 g/l and 0.28 g/l respectively. Estimate the retention time of the solute at a flow rate of 2 l/min.

9.2. A chromatogram was obtained with an amino acid using a 30 cm long gel-filtration column. It was found that: $t_M = 2$ minutes; $t_0 = 5$ minutes; $C_0 = 5$ g/l; $w = 4$ min; voidage fraction = 0.3. If the effluent between 4 to 6 minutes were collected, what fraction of the amino acid would be recovered?

9.3. An antibody was injected into an ion-exchange column and the following data was obtained from the chromatogram:
$C_0 = 1$ kg/m^3, $t_0 = 5$ minutes, $\sigma = 1$
Predict the time at which the concentration of the antibody in the effluent stream reaches 0.5 kg/m^3.

9.4. The van Deemter plot for an antibiotic obtained with a 1 m long chromatographic column having a voidage fraction of 0.25 is shown in the figure below. Predict the number of theoretical plates obtained with this column at the optimum mobile phase velocity. If the retention time of the antibiotic at the optimum mobile phase velocity is 3 minutes, predict its characteristic peak width. With the same column at the same mobile phase velocity, the retention time of an impurity was found to be 4 minutes. Predict the resolution and selectivity.

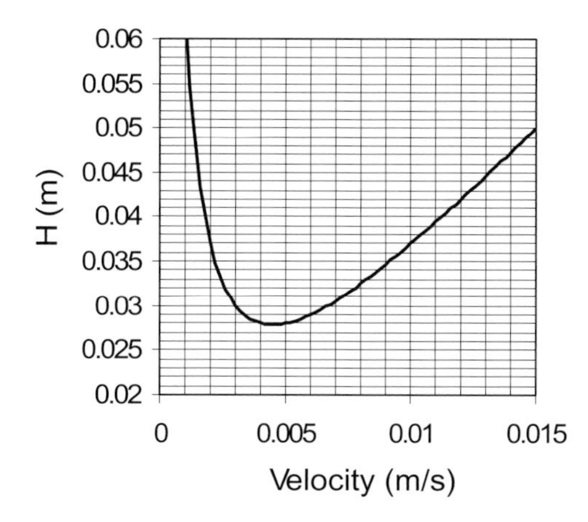

9.5. The solute concentration in a chromatographic peak is give by the equation shown below:

$$C = 2\exp\left(-\frac{(t-10)^2}{2}\right)$$

In this equation C is in g/l and t is in minutes. When will the solute concentration in the peak be equal to 1 g/l?

9.6. The chromatogram of a solute obtained using a 30 cm long chromatographic column is shown in the figure below. Calculate the height equivalent of a theoretical plate for this column. If the mobile phase retention time at the same mobile phase flow rate is 1.5 minutes, calculate the capacity factor for the solute. If the voidage fraction of the column is 0.25, calculate the distribution coefficient. If effluent between 1 and 5 minutes is collected, what percent of the solute will be obtained?

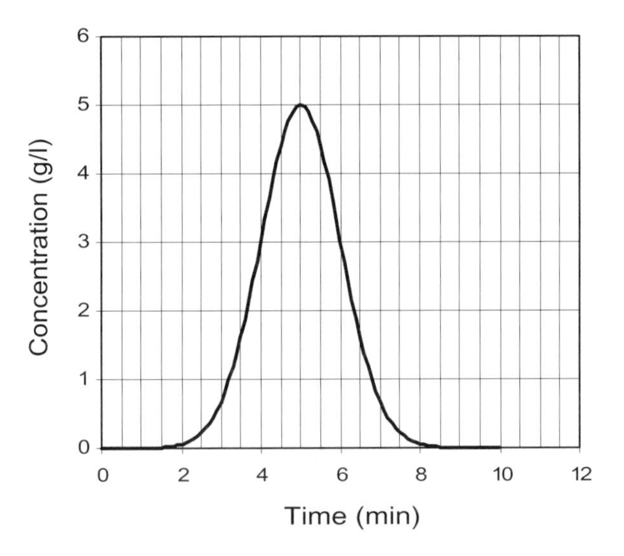

9.7. Three whey proteins (A, B and C) are being fractionated by isocratic chromatography using a 15 cm long cation exchange column having 90 theoretical plates. The column has a voidage fraction of 0.35 and a diameter of 9 mm. A mobile phase flow rate of 1.5 ml/min resulted in retention times of 10, 18 and 23 minutes respectively for proteins A, B and C. Comment on the feasibility of separation.

References

P.A. Belter, E.L. Cussler, W.-S. Hu, Bioseparations: Downstream Processing for Biotechnology, John Wiley and Sons, New York (1988).

M.R. Ladisch, Bioseparations Engineering: Principles, Practice and Economics, John Wiley and Sons, New York (2001).

J.M. Coulson, J.F. Richardson, J.R. Backhurst, J.H. Harker, Coulson and Richardson's Chemical Engineering vol 2, 4th edition, Butterworth-Heinemann, Oxford (1991).

L.R. Snyder, J.J. Kirkland, Introduction to Modern Liquid Chromatography, Wiley Interscience, New York (1979).

J.M. Miller, Chromatography: Concepts and Contrasts, Wiley Interscience, New York (2004).

E. Klein, Affinity Membranes: Their Chemistry and Performance in Adsorptive Separation Processes, Wiley Interscience, New York (1991).

Chapter 10

Filtration

10.1. Introduction

Filtration is a separation process in which a solid-liquid mixture called the feed (or the suspension) is forced through a porous medium on which the solids are deposited or in which they are entrapped. The porous medium which allows the liquid to go through while retaining the solids is called the filter. The retained solid is called "the residue" or "the cake". The clarified liquid is called "the effluent" or the "filtrate".

Filtration can be broadly classified into three categories (see Fig. 10.1). If recovery of solids from high solid content slurry is desired, the process is called cake filtration. The term clarification is applied when the solid content in the feed does not exceed 1 wt %. In a clarification process the filtrate is the primary product. The third type of filtration is called cross-flow filtration in which the liquid flows parallel to the filtration medium. Cross-flow filtration is mainly used for membrane filtration and will be discussed in details in the chapter on membrane based bioseparation processes. In this chapter cake filtration will mainly be discussed.

Some of the applications of filtration in the bio-industry are:

1. Recovery of crystalline solids
2. Recovery of cells from fermentation medium
3. Clarification of liquids and gases
4. Sterilisation of liquids

10.2. Theory of filtration

There are two main mechanisms by which solids are retained by a filter (see Fig. 10.2):

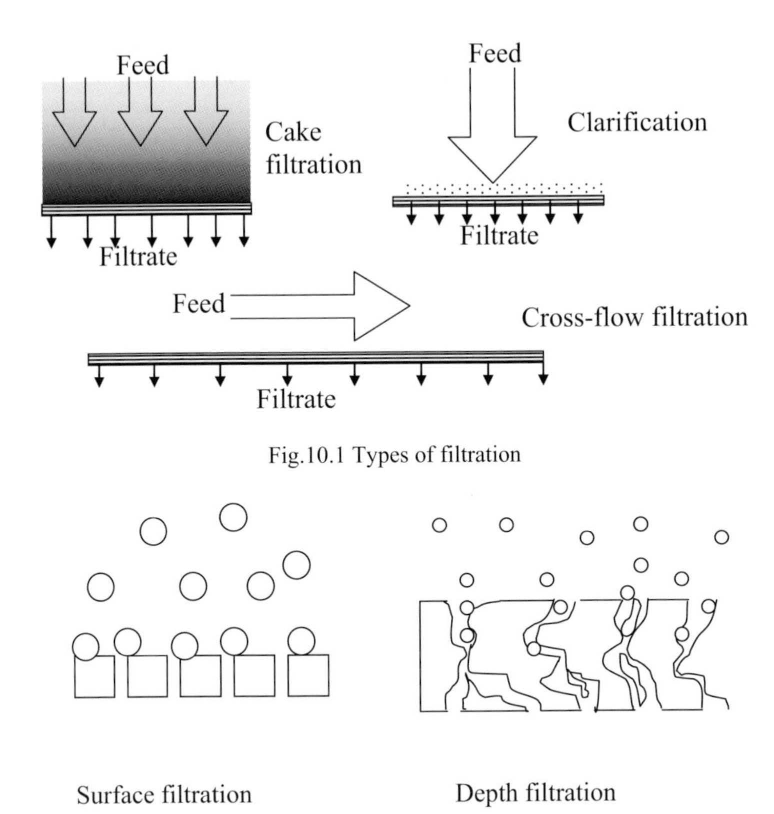

Fig.10.1 Types of filtration

Fig. 10.2 Mechanisms of filtration

Surface filtration: The particles are retained by a screening action and held on the external surface of the filter. The particles are not allowed to enter the filtration medium. Cake filtration and cross-flow filtration are based on surface filtration.

Depth filtration: Particles are allowed to penetrate pores and pore networks present in the filtration medium. They are retained within the filter by three mechanisms: direct interception, inertial impaction and diffusional interception. In direct interception the particles enter the pores or pore networks within the filtration medium and get trapped where the pore diameter becomes equal to the particle diameter. Particles whose diameters are significantly smaller than the pore diameter get trapped where pores are already constricted by collected particles. Particles being carried by fluids possess momentum on account of their mass and velocity. Pores present in most filtration media are tortuous in nature. The flow of fluids through these pores is usually laminar in nature on account of their small diameters. At tortuous sections of pores the fluid streamlines follow the curves while particles due to their inertia continue to move straight and as a result hit the filter medium, lose their momentum and are retained. This collection mechanism which is referred to as inertial impaction is important for particles larger than 1.0 μm in diameter. Diffusional interception is more relevant to filtration of gases. The gas molecules, due to their random motion continually bombard suspended particles, particularly those smaller than 0.3 μm in diameter. The suspended particles therefore deviate from their streamlines and impact on the filter medium. Once this happens, the particles lose their momentum and are retained. Most clarification processes rely of depth filtration.

10.3. Filter medium

The function of a filter medium is primarily to act as an impermeable barrier for particulate matter. In clarification processes the filtration medium is usually the only barrier present. At the beginning of a cake filtration process, the role of the filter medium is to act as a barrier. However, once the cake formation commences, the cake becomes the main particle-retaining barrier and the role of the filter medium is mainly as a support for the cake. The filter medium should have sufficient mechanical strength, should be resistant to corrosive action of fluids being processed and should offer low resistance to the flow of filtrate. Commonly used filter media are:

1. Filter paper
2. Woven material (e.g. cheese cloth, woven polymer fibre, woven glass fibre)
3. Non-woven fibre pads
4. Sintered and perforated glass
5. Sintered and perforated metal
6. Ceramics
7. Synthetic membranes

10.4. Driving force

Filtration is driven by applying a pressure drop across the filter medium. The driving force can be applied by pressurizing the feed side (i.e. positive pressure filtration or simply pressure filtration) or by creating a vacuum in the filtrate side (i.e. negative pressure filtration or vacuum filtration). These two types of filtration are shown in Fig. 10.3. In the industry, both pressure and vacuum filtration are used. Pressure filtration can be driven by pressurizing the feed using compressed air pressurisation or by maintaining a hydrostatic liquid head on the feed side. The feed can also be pressurized with a suitable pump. Vacuum filtration is commonly used in the laboratory since pressure vessels are not required. Vacuum can also be easily generated using a water jet injector or a vacuum pump. Vacuum filtration is preferred from a safety point of view since explosion is more hazardous than implosion. However, with vacuum filtration the maximum pressure drop is restricted to 1 atmosphere. Also substances that form foam cannot be filtered by vacuum filtration.

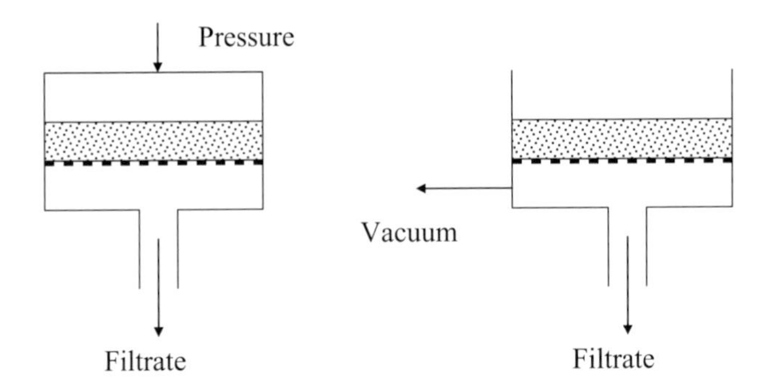

Fig. 10.3 Pressure and vacuum filtration

10.5. Constant pressure cake filtration

Constant pressure filtration refers to a filtration process where the driving force (i.e. the pressure drop across the filter medium) is kept constant. If we consider the filtration of a Newtonian liquid, which is free from particles, the flow of liquid through the filter can be explained by Darcy's law:

$$Q = \frac{kA\Delta P}{\mu l} \tag{10.1}$$

Where

Q = volumetric filtration rate (m^3/s)

k = Darcy's law permeability (m^2)
A = area of filter medium (m^2)
ΔP = pressure drop across the filter medium (Pa)
μ = viscosity (kg/m s)
l = thickness of the filter medium (m)

The permeability and thickness of a filter can be combined into a medium resistance term and equation (10.1) can be written as:

$$Q = \frac{A\Delta P}{\mu R_M} \tag{10.2}$$

Where
R_M = media resistance (/m)

The filtration rate can be expressed in terms of the volume of filtrate collected as shown below:

$$Q = \frac{dV(t)}{dt} = \frac{A\Delta P}{\mu R_M} \tag{10.3}$$

Where
$V(t)$ = cumulative filtrate volume (m^3)

The value of R_M can be determined from a plot of $V(t)$ versus t (see Fig. 10.4).

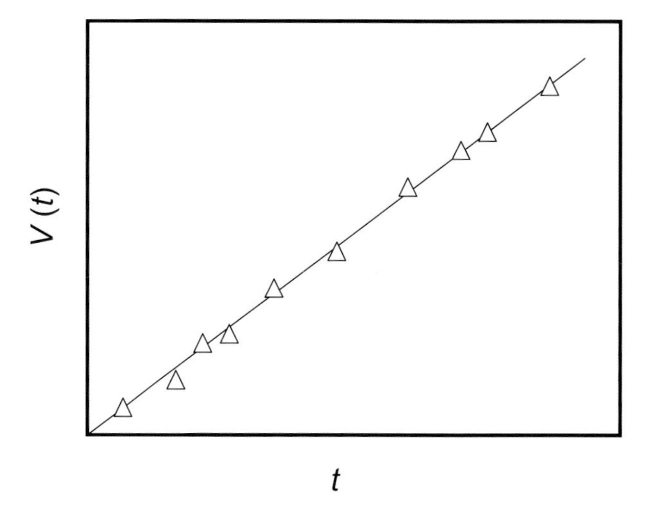

Fig. 10.4 Determination of medium resistance

Example 10.1

Water is being filtered at 25°C through a filter paper disc having a diameter of 10 cm using a pressure drop of 20 kPa. The filtration rate is found to be 5 ml/min. Predict the filtration rate if a stack of 5 filter papers were to be used as the filtration medium, the pressure drop remaining the same.

Solution

Water filtration through filter paper can be explained by Darcy law. Therefore:

$$Q \propto \frac{1}{l}$$

Therefore:

$$\frac{Q_1}{Q_2} = \frac{l_2}{l_1}$$

It is given that:

$Q_1 = 5$ ml/min

$l_2 = 5l_1$

Therefore:

$$Q_2 = \frac{5}{5} \text{ ml/min} = 1 \text{ ml/min}$$

When the substance being filtered contains retained suspended particles these particles from a cake layer on top of the filter medium and offer added resistance to flow of filtrate. For cake filtration, equation (10.3) can be written as:

$$\frac{dV(t)}{dt} = \frac{A\Delta P}{\mu(R_M + R_c(t))} \tag{10.4}$$

Where

$R_C(t)$ = cake resistance (/m)

In cake filtration particles are continuously deposited on the filter medium and hence the thickness of the cake layer increases. Therefore the cake resistance is also increases continuously. The thickness of cake deposited is proportional to the cumulative amount of feed filtered and hence on the cumulative volume of filtrate. The cake thickness is inversely proportional to the filter area. Therefore, we can write:

$$\frac{dV(t)}{dt} = \frac{A\Delta P}{\mu\left(R_M + \alpha C_s\left(\frac{V(t)}{A}\right)\right)} \tag{10.5}$$

Where

α = specific cake resistance (m/kg)

C_s = mass of cake solids per unit volume of filtrate (kg/m^3)

If this differential equation is solved with the initial condition: $V(t) = 0$ at $t = 0$, we get:

$$t = \frac{\mu}{A\Delta P}\left(\frac{\alpha C_s V(t)^2}{2A} + R_M V(t)\right) \tag{10.6}$$

Equation (10.6) can be written as:

$$t = K_p V(t)^2 + BV(t) \tag{10.7}$$

Where

K_p = $(\mu\, C_s\, \alpha\, /\, 2\, \Delta P\, A^2)$ = cake constant (s/m^6)

B = $(\mu\, R_M\, /\, \Delta P\, A)$ = media constant (s/m^3)

Equation (10.7) can be written as:

$$\frac{t}{V(t)} = K_p V(t) + B \tag{10.8}$$

A plot of $[t\, /\, V(t)]$ versus $V(t)$ gives a straight line from which K_p (the slope) and B (the intercept) can easily be determined (see Fig. 10.5). From the experimentally determined values of K_p and B, the specific cake resistance and medium resistance can be determined. In a cake filtration process where a significant amount of cake is allowed to accumulate, the medium resistance can become negligible compared with the cake resistance. In such situations, equation (10.7) can be written as:

$$t = \frac{\mu \alpha c_s}{2A^2\Delta P} V(t)^2 \tag{10.9}$$

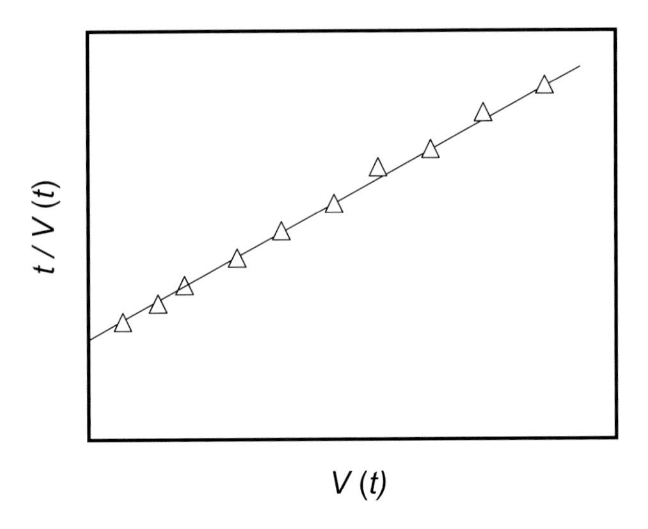

Fig. 10.5 Cake filtration plot

Example 10.2

The following data was obtained from a constant pressure cake filtration experiment:

Time (s)	5	10	20	30
V(t) (litres)	0.040	0.055	0.080	0.095

The following additional information is given:
$A = 0.1$ ft^2, $C_s = 0.015$ kg/l, $\mu = 1.1$ centipoise, $\Delta P = 10$ N/m^2.

a) Determine R_M
b) Determine the specific cake resistance

Solution

This problem can be solved using equation (10.8). From a plot of $(t/V(t))$ versus $V(t)$ the values of K_p (slope) and B (intercept) can be determined.

$$B = \frac{\mu R_M}{\Delta P A} = 0$$

Therefore:

$R_M = 0$

$$K_p = \frac{\mu C_S \alpha}{2\Delta P A^2} = 3 \times 10^9$$

$\mu = 1.1\,\text{cp} = 1.1 \times 10^{-3}$ kg/m s
$C_S = 0.015$ kg/l $= 0.015 \times 10^{-3}$ kg/m^3
$A = 0.1$ ft$^2 = 9.29 \times 10^{-3}$ m^2
Therefore:
$\alpha = 3.139 \times 10^8$ m/kg

Example 10.3
50 litres of filtrate is collected in 30 minutes when an inorganic suspension is filtered through a sintered glass filter using a pressure drop of 50 kPa. How much filtrate will be collected in 30 minutes at a pressure drop of 100 kPa? Assume that R_M is negligible.

Solution
This problem can be solved using equation (10.9) since the medium resistance is negligible. From equation (10.9) we can write:

$$V(t) = \sqrt{\frac{2tA^2\Delta P}{\mu\alpha C_S}}$$

Since t, A, μ, α and C_S remains the same, we can write:

$$\frac{V(t)_1}{V(t)_2} = \sqrt{\frac{\Delta P_1}{\Delta P_2}}$$

It is given that:
$V_1 = 50$ l
$\Delta P_1 = 50$ kPa
$\Delta P_2 = 100$ kPa
Therefore:

$$V(t)_2 = 50 \times \sqrt{\frac{100}{50}} = 70.711$$

So far the cake has been considered to have uniform properties during the entire filtration process. This assumption is valid when the cake material in incompressible, i.e. its specific resistance is independent of the applied pressure. Most crystalline materials form incompressible cakes. However, some substances form cakes, which deform due to application of pressure and are hence compressible. Examples include yeast cells, mould cells and precipitated proteins. Different types of correlation are available for compressible cakes. A typical form is shown below:

$$\alpha = \alpha'(\Delta P / \Delta P_0)^s \quad (10.10)$$

Where

α' = cake constant (m/kg)

s = cake compressibility factor (-)

ΔP_0 = dimension correcting factor (Pa)

The cake constant depends on the size and shape of particles. The cake compressibility factor varies between 0.1 and 0.8. For an incompressible cake $s = 0$.

10.6. Constant rate cake filtration

Constant rate filtration refers to a filtration process where the filtration rate is kept constant by appropriately adjusting the pressure drop during the process. When the filtration rate (Q) is constant, equation (10.5) can be written as:

$$Q = \frac{A \Delta P(t)}{\mu \left(R_M + \frac{\alpha C_s Q t}{A} \right)} \quad (10.11)$$

Where

$\Delta P\ (t)$ = instantaneous pressure drop (Pa)

Rearranging equation (10.11) we get:

$$\Delta P(t) = \frac{\mu \alpha C_s Q^2 t}{A^2} + \frac{\mu Q R_M}{A} \quad (10.12)$$

When the medium resistance is negligible compared to the cake resistance, we can write equation (10.12) as:

$$\Delta P(t) = \frac{\mu \alpha C_s Q^2 t}{A^2} \quad (10.13)$$

Example 10.4

Time-pressure drop data for a constant rate cake filtration process is given in the table below. If we are to carry out the filtration at twice the filtration rate, predict the pressure drop required after 1 hour of filtration. Assume that the cake is incompressible.

Time (min)	0	10	20	30	40	50
ΔP (kPa)	5	10	15	20	25	30

Solution

This problem can be solved using equation (10.12). If the experimental data shown above is plotted (i.e. ΔP (t) versus t)), the best fitting straight line is of the form:

$$\Delta P(t) = 0.5t + 5 \qquad (10.a)$$

Where ΔP (t) is in kPa and t is in minutes. This apparent use of inconsistent units is not a problem here due to the manner in which this problem is going to be solved. Where μ, α, C_S, A and R_M remain the same, the slope of equation (10.12) is proportional to the square of Q and the intercept is proportional to Q. Therefore when the filtration rate (Q) is doubled, slope becomes four times and the intercept becomes twice. Therefore the equation for twice the filtration rate is:

$$\Delta P(t) = 2t + 10 \qquad (10.b)$$

Using equation (10.b), the pressure drop after 60 minutes can be determined:

$$\Delta P(t) = ((2 \times 60) + 10)\text{ kPa} = 130\text{ kPa}$$

10.7. Improvement of filtration efficiency

The efficiency of cake filtration depends on the achieving high cake accumulation on the filter medium. However, the filtration rate declines with cake accumulation due to the increase in cake resistance. One way to solve this problem is to alter cake properties such that the specific cake resistance is reduced. This can be achieved by:

Feed pre-treatment

The feed can be pre-treated by physical methods (e.g. heating) or by addition of chemicals (e.g. coagulants, flocculants) to obtain a porous cake with low specific cake resistance. However, thermolabile substance cannot be heated. Moreover addition of coagulants and flocculants might not be possible in some applications.

Filter aids

Filter aids are substances that are mixed with the feed for creating very porous cakes. This increases the filtration rate very significantly. The filter aids which are particulate in nature can later be removed from the dried and powdered cake by suitable separation techniques (e.g.

sieving). However, in certain cases it might not be possible to completely remove the filter aid and their use is restricted. Filter aids are rarely used when the cake is the product of interest. Certain deformable substances present in the feed can block the pores within the filtration medium. When such substances are being filtered, it might be a good idea to precoat the medium with a layer of filter aid.

10.8. Mode of operation

Filtration can be carried out in different ways:

Cake accumulation and removal in batch mode

This is the commonest mode for small-scale cake filtration. A batch of feed is pumped into the filter unit and filtration is carried out either at constant rate or at constant pressure. The process is terminated when the filtration rate gets unacceptably low, or when the pressure required gets too high, or when the filtration device is filled with the filter cake. The cake is then removed from the device by scraping it off the filter medium. Often this requires dismantling of the filtration unit. The filter medium is usually then cleaned and made ready for the next batch. The general scheme for this mode of operation is shown in Fig. 10.6. Examples of filtration devices operated in this mode include:

1. Funnel filter
2. Filter press
3. Leaf pressure filter
4. Vacuum leaf filter

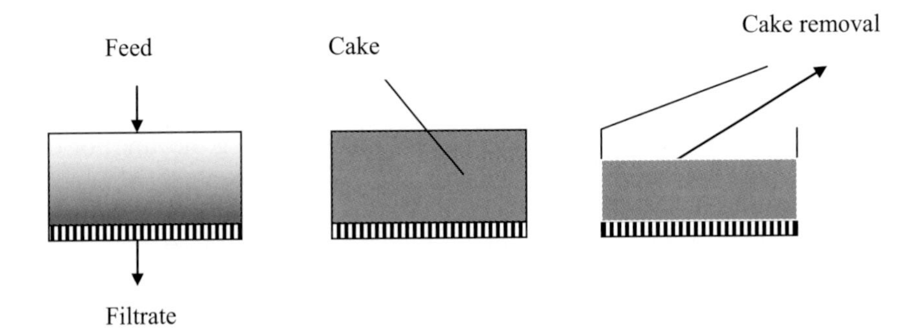

Fig. 10.6 Batch cake filtration

Continuous cake accumulation and removal

A batch process may not be suitable for large-scale cake filtration. Continuous filters which allow simultaneous cake accumulation and removal are usually used for large-scale processes. Examples of such devices include:

1. Horizontal continuous filter
2. Rotary drum filter

Slurry concentration by delayed cake filtration

When the objective of the filtration process is to thicken the slurry, the build-up of cake on the filter medium is avoided. Slurry thickening can be achieved by controlling the thickness of the cake layer thereby keeping the particulate matter in a suspended form on the feed side. This can be achieved by incorporating mechanical devices such as moving blades, which continuously scrape of the cake from the filter surface. With the moving blade arrangement, the thickness of the cake is limited by the clearance between the filter medium and the blade. This type of filtration can be carried out until the solid content on the feed side reaches a critical level beyond which the slurry does not flow.

Slurry concentration by cross-flow filtration

An alternative way by which the build-up of cake on the filter can be discouraged is by using a cross-flow mode of operation. This is achieved by maintaining a very high velocity of feed flow parallel to the surface of the filter medium. Typical cross-flow rates may be 10 – 20 time the filtration rate. However, cross flow filtration cannot be used for obtaining very thick slurry since the energy required for maintaining the cross-flow velocity becomes prohibitively high.

Cake washing

After its formation, the cake may contain a significant amount of entrapped liquid. When the liquid is the product of interest, this entrapment represents a loss of yield. When the cake is itself the product, the entrapped liquid represents the presence of impurity. The entrapped liquid can be removed by cake washing. The washing liquid should itself not be a "new impurity". Its presence should either be acceptable in the final product (i.e. either the cake or the filtrate), or it should be easily "removable". If the product is soluble in the washing liquid, as often is the case, the duration of the washing process depends on a trade-off between the amount lost and the purity desired.

10.9. Filtration equipment

Filter Press

A filter press consists of a series of horizontally arranged vertical filter elements, each consisting of a frame within which the cake can be accumulated sandwiched between filter medium on either side. Each filter medium is supported on a plate which has grooves to allow easy collection of the filtrate. The "press" refers to the external structure which provides the necessary force to seal each filter element and supports the plate and frame assembly. Fig. 10.7 shows the individual plates used in a filter press. Both faces of each plate are covered with filter medium and together with the frames these form a series of chambers into which the feed is introduced under pressure. The filter medium retains suspended solids and the filter cake builds up within the chamber. When the filtration cycle is complete the pressure holding the system together is released and the filter plates are separated to remove the cake from within the frames. The operation of a plate and frame filter press is summarized in Fig. 10.8.

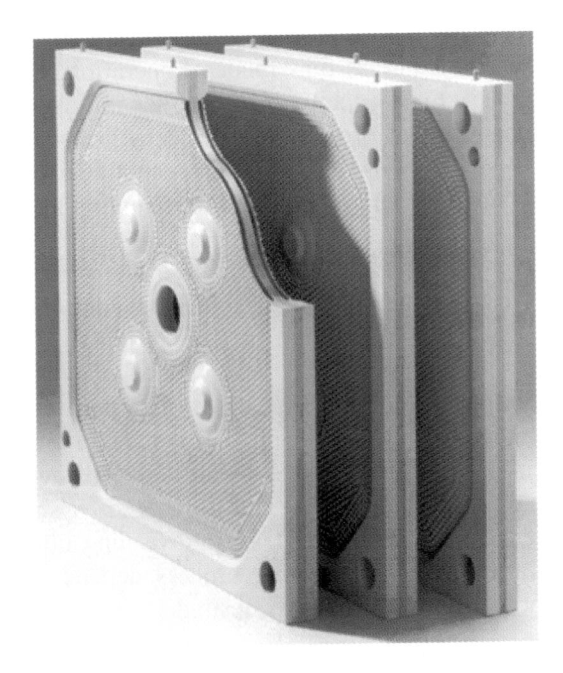

Fig. 10.7 Filter press plates (Photo courtesy of Menardi)

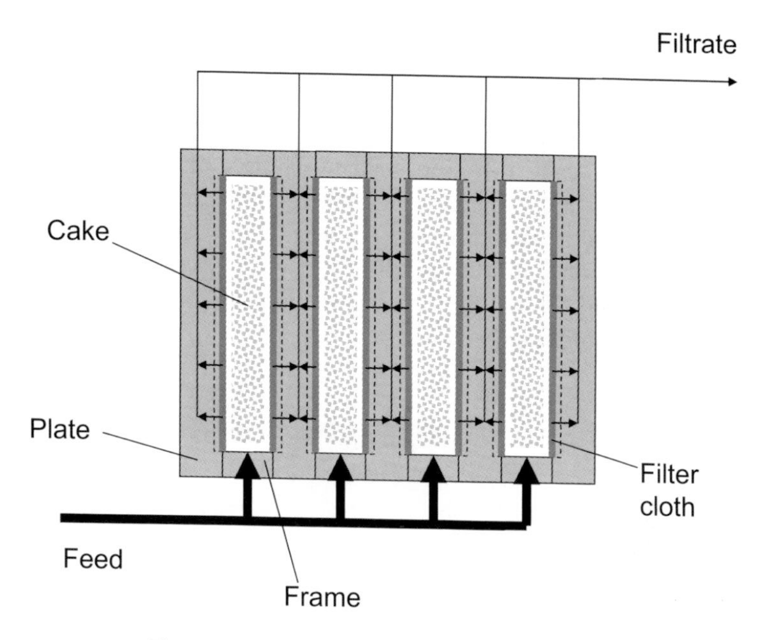

Fig. 10.8 Operation of plate and frame filter press

A plate and frame filter press is mainly used for cake filtration and cake washing. It could also be used for an extended clarification process but this is rarely done. It is frequently used for solid-liquid extraction (or leaching). It is a common sight in the chemical, pharmaceutical, food, metallurgical and ceramic industries. The main advantages of this device are its compact design and its high throughput. Disadvantages include high labor cost (due to the need for assembling and subsequent dismantling), high down times, capacity limitation and batch-wise operation.

Rotary Drum Vacuum Filter

A rotary drum vacuum filter is a continuous filtration device in which the solids are separated by a porous filter cloth or similar filtration media wrapped around a drum-like structure with a perforated curved surface. The drum is rotated, partially submerged, through a feed solution held in a trough and vacuum applied on the inside. The filtrate flows through the filter media to the inside of the drum and the cake accumulates on the outside. The working principle of a rotary drum filter is shown in Fig. 10.9. At any given location on the filter, the cake layer builds up as it moves through the feed. As this emerges from the feed, the vacuum continues to draw the liquid from the cake, dewatering it in the process. If required, the cake can be washed by spraying it with a wash liquid and further dewatered. The semi-dry cake is then removed from the filter medium by using a fixed knife or a cutting wire. A rotary vacuum filter is mainly used for cake filtration, cake washing and de-watering in the chemical pharmaceutical, metallurgical and ceramic industries. These are also used for municipal wastewater treatment. Advantages include continuous operation and the possibility of cake drying. Disadvantages include low driving force (due to being vacuum driven) and complicated design with many moving parts and seals.

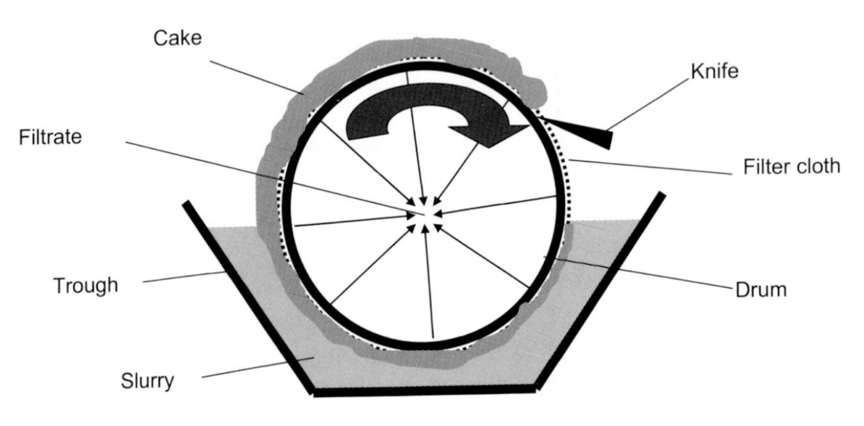

Fig. 10.9 The working principle of a rotary drum filter

Pressure leaf filter

A pressure leaf filter consists of a number of rectangular basic filtration units (also called leaves) connected together in parallel by means of flexible hose or a rigid tube manifold (Fig. 10.10). Each leaf is made up of a light metal frame made from wire mesh. These leaves are covered with filter cloth or woven wire cloth. The leaf assembly is housed within a pressure vessel into which the feed is pumped at high pressure. The filtration is driven by pressure and the cake is accumulated within the pressure vessel. A pressure leaf filter may be used for cake filtration and clarification. Advantages include simplicity of design and flexibility of use. Disadvantages include high labor cost and high equipment footprint.

Fig. 10.10 Pressure leaf filter (Photo courtesy of Menardi)

Exercise problems

10.1. A suspension of yeast cells is to be filtered at a constant filtration rate of 50 l/min. The suspension has a solid content of 70 kg/m^3 of suspension and the yeast cells have a bulk density of 800 kg/m^3. Laboratory tests indicate that the specific cake resistance is 40 m/kg and the viscosity of the filtrate is 2.9×10^{-3} kg/m s. The filter has an area of 0.1 m^2 and the medium offers negligible resistance. How long can the filtration rate be maintained before the pressure drop exceeds 1 N/m^2? What volume of cake and filtrate are collected during this time? (note that in the cake filtration equations, c_s is expressed in mass of solids per volume of filtrate and NOT total volume of suspension)

10.2. Pure ethanol is being filtered through a porous disc having a diameter of 2 cm and a thickness of 1 mm. When a pressure drop of 100 kPa was used to drive the filtration, the filtration rate obtained was 10 ml/min. If a porous disc made of the same material but having a diameter of 5 cm and a thickness of 3 mm were used for filtering ethanol, what filtration rate would you expect at 150 kPa pressure drop?

10.3. The data obtained from a constant pressure cake filtration experiment is shown below:

Time (s)	5	10	20	30
$V(t)$ (litres)	0.040	0.055	0.080	0.095

Predict the cumulative volume of filtrate collected after 4 minutes of filtration. What would have been the cumulative filtrate volume collected after 30 seconds had the solids content in the feed been double of that used in this experiment?

10.4. When an aspirin suspension was filtered through a porous glass disc having a diameter of 0.1 m at a constant rate of 4×10^{-5} m^3/s, the pressure drop driving filtration was found to be 10,000 Pa after 2000 seconds of filtration. If the same suspension were filtered at a rate of 4×10^{-4} m^3/s using another porous glass disc having a diameter of 0.25 m, what would the pressure drop be after 5000 seconds of filtration? Assume that the media resistance is negligible in both cases.

References

P.A. Belter, E.L. Cussler, W.-S. Hu, Bioseparations: Downstream Processing for Biotechnology, John Wiley and Sons, New York (1988).

J.M. Coulson, J.F. Richardson, J.R. Backhurst, J.H. Harker, Coulson and Richardson's Chemical Engineering vol 2, 4th edition, Butterworth-Heinemann, Oxford (1991).

C. J. Geankoplis, Transport Processes and Separation Process Principles, 4th edition, Prentice Hall, Upper Saddle River (2003).

W.L. McCabe, J.C. Smith, P. Harriott, Unit Operations of Chemical Engineering, 7th edition, McGraw Hill, New York (2005).

Chapter 11

Membrane based bioseparation

11.1. Introduction

A membrane is a thin semi-permeable barrier which can be used for the following types of separation:

1. Particle-liquid separation
2. Particle-solute separation
3. Solute-solvent separation
4. Solute-solute separation

Among the many applications are: product concentration, product sterilization (i.e. removal of bacteria and virus particles), solute fractionation, solute removal from solutions (e.g. desalination, demineralization), purification, and clarification. Some of the factors, which are utilized in membrane based separation, are:

1. Solute size
2. Electrostatic charge
3. Diffusivity
4. Solute shape

The transport of material through a membrane could be driven by convection or by diffusion or indeed by a combination of the two. Convection based transport takes place due to transmembrane pressure and hence membrane processes involving convective transport are also referred to as pressure driven processes. The manner in which a pressure driven separation process is carried out is shown in Fig. 11.1. Diffusion based transport utilizes the concentration difference of the transported species across the membrane as the driving force. The manner in which a diffusion driven separation process is carried out is shown in Fig. 11.2.

A membrane may be made from organic polymers or inorganic material such as glass, metals and ceramics, or even liquids. Examples of polymeric (or organic) membranes include those made from cellulose,

cellulose acetate (CA), polysulfone (PS), polyethersulfone (PES), polyamides (PA), polyvinylidine fluoride (PVDF), polyacrylonitrile (PAN). Inorganic membranes can be made from ceramics, glass, pyrolyzed carbon and stainless steel.

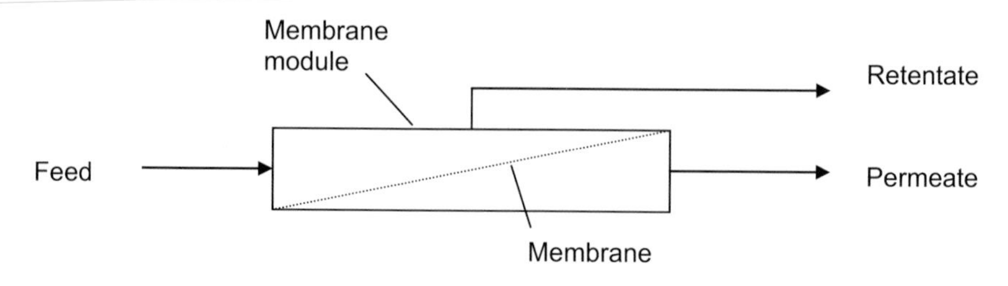

Fig. 11.1 Pressure driven separation

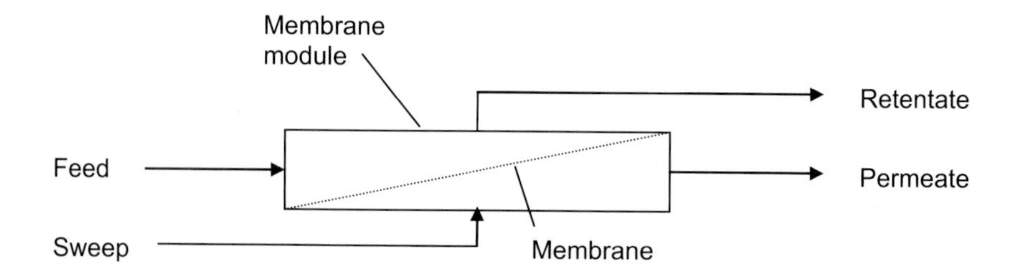

Fig. 11.2 Diffusion driven separation

From a structural point of view membranes are broadly divided into two types as shown in Fig. 11.3:

1. Symmetric (or isotropic)
2. Asymmetric (or anisotropic)

A symmetric membrane has similar structural composition and morphology at all positions within it. An asymmetric membrane is composed of two or more structural planes of non-identical composition or morphology. From a morphological point of view, membranes can be classified into two categories: porous or dense. Porous membranes have tiny pores or pore networks (see Fig. 11.4). On the other hand dense membranes do not have any pores and solute or solvent transport through these membranes take place by a partition-diffusion-partition mechanism.

Membranes are available in three basic forms:

1. Flat sheet membrane
2. Tubular membrane
3. Hollow fibre membrane

Symmetric membranes

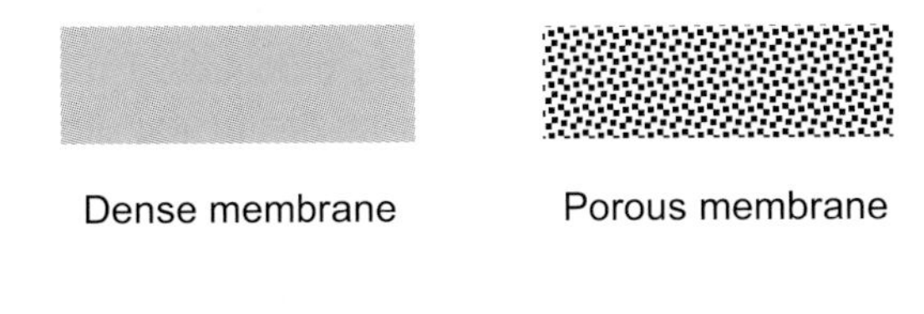

Asymmetric membranes

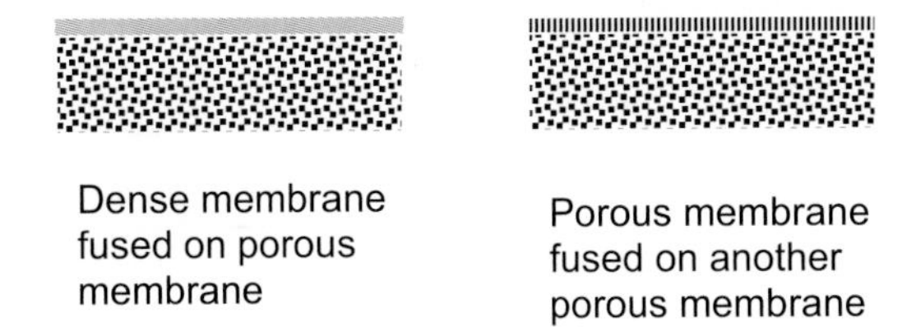

Fig. 11.3 Symmetric and asymmetric membranes

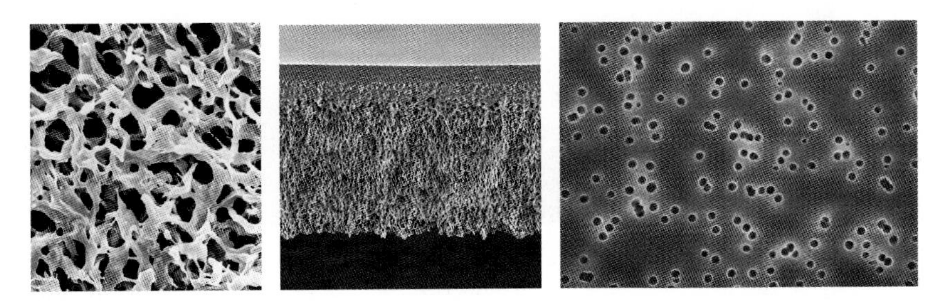

Fig. 11.4 Porous membranes (A. Microporous symmetric membrane, B. Microporous asymmetric membrane, C. Isoporous membrane) (Micrographs courtesy of Millipore Corporation)

Flat sheet membranes look like filter paper (see Fig. 11.5). They are available is the form of filter discs or rectangular sheets. The most common type of tubular membrane looks like a single hollow tube of circular cross-section, the wall of the tube functioning as the membrane. Tubular membranes having square and other types of cross-section are also available. Monolith tubular membranes look like cylindrical blocks with large numbers of parallel tubes within them. These tubes typically have diameter in the range 0.5 cm to 2 cm. Fig. 11.6 shows how a tubular membrane is used. Hollow fibres are also tube-like in appearance. However, these membranes have much smaller diameters than tubular membranes. Typical fibre diameter is of the order or 1 mm. Fig. 11.7 shows hollow fibre membranes of different diameters. If the inner wall of hollow fibre acts as the membrane, it is of the inside-out type whereas when the outer wall acts as the membrane, it is referred to as the outside-in type. Double skinned hollow fibre membranes which can function both as inside-out and outside-in membranes are also available.

Fig. 11.5 Flat sheet membranes

The performance of a membrane depends to a large extent on the properties of the membrane. Thus membrane characterization is important both for membrane makers and membrane users. Some of the properties which need characterizing are:

1. Mechanical strength e.g. tensile strength, bursting pressure
2. Chemical resistance e.g. pH range, solvent compatibility
3. Permeability to different species e.g. pure water permeability, sieving coefficient
4. Average porosity and pore size distribution

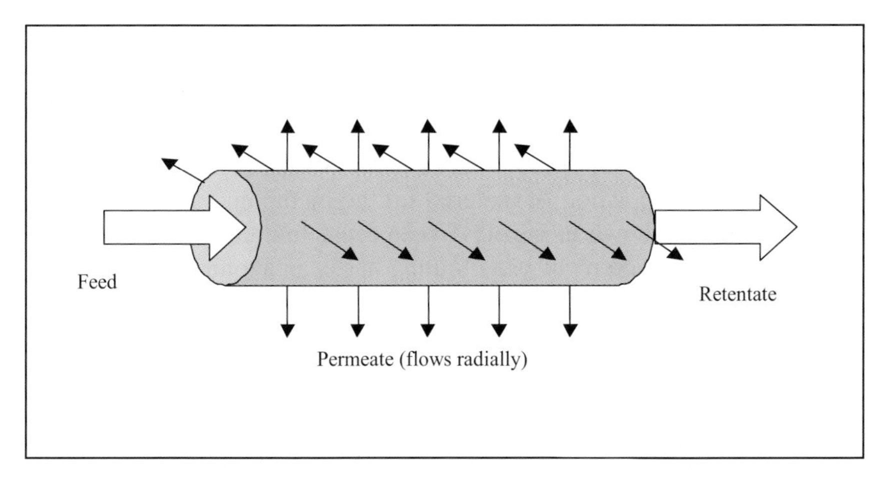

Fig. 11.6 Tubular membrane

Fig. 11.7 Hollow fibre membranes

The throughput of material through a membrane is frequently referred to as the flux. The flux depends on the applied driving force e.g. transmembrane pressure or concentration gradient as well as on the resistance offered by the membrane. The decline in flux through a membrane with time in a constant driving force membrane process is generally due to fouling (see Fig. 11.8). Fouling refers to an increase in membrane resistance during a process. Many membrane processes are operated at constant flux. In constant flux membrane processes, fouling manifests itself in terms of the increased driving force (e.g. pressure) required to sustain a particular flux through the membrane (see Fig. 11.9). Fouling is an undesirable phenomenon which is usually caused by adsorption and deposition of material on the membrane. One or both of two mechanisms shown in Fig. 11.10 can cause fouling. In this chapter, a detailed discussion on membrane fouling has been avoided since this is a very case dependent phenomenon.

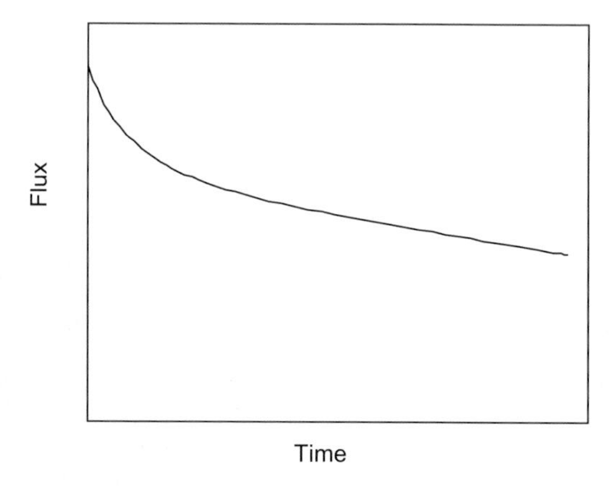

Fig. 11.8 Decline in flux due to fouling in constant driving force membrane separation

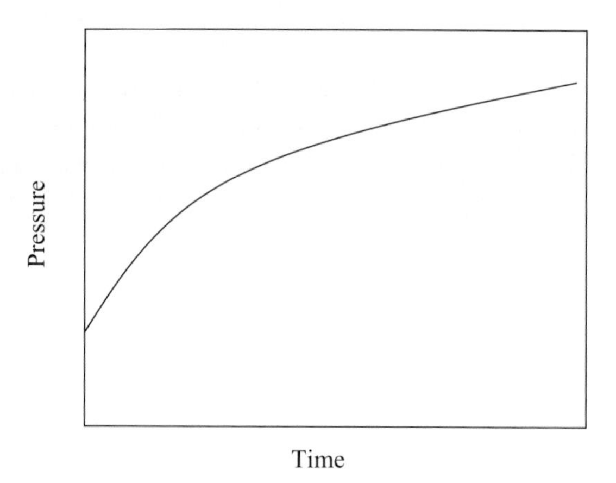

Fig. 11.9 Fouling in constant flux membrane separation

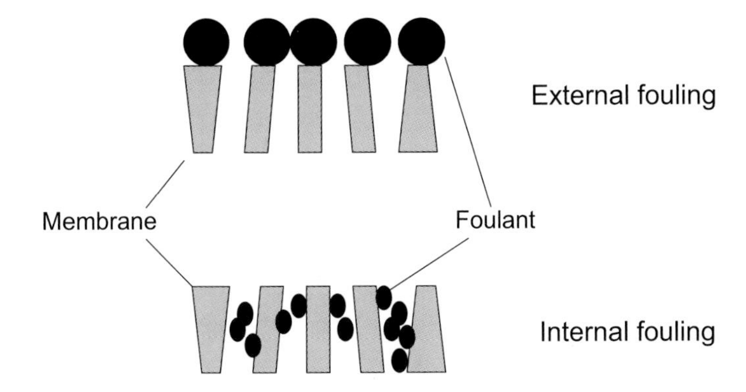

Fig. 11.10 Mechanisms of membrane fouling

11.2. Classification of membrane processes

Pressure driven membrane based bioseparation processes can be classified into four types based on the size of the permeable species (see Fig. 11.11). A fifth type called dialysis allows solutes similar those in nanofiltration to pass through. However, unlike nanofiltration, which is a pressure driven process, dialysis is a concentration gradient (or diffusion) driven process.

Microfiltration (MF) is used for separation of fine particles from solutions. The transmembrane pressure ranges from 1 to 50 psig. Most microfiltration membranes capture particles by surface filtration, i.e. on the surface of the membrane. In some cases depth filtration is also used. MF is most commonly used for clarification, sterilization and slurry concentration. Most MF membranes are symmetric.

Ultrafiltration (UF) membranes retain macromolecules such as proteins while allowing smaller molecules to pass through. UF is used to (a) separate large molecules from solvents, (b) separate large molecules from smaller molecules, and (c) separate large molecules from one another. The primary separation mechanism is size exclusion, but physicochemical interactions between the solutes and the membrane, and operating conditions can influence the process quite significantly. Normal transmembrane pressure in ultrafiltration ranges from 10 to 100 psig. Most UF membranes are asymmetric.

Nanofiltration (NF) membranes allow salts and other small molecules to pass through but retain larger molecules such as peptides, hormones and sugars. The transmembrane pressure in NF ranges from 40 to 200 psig. Most NF membranes are composite i.e. asymmetric.

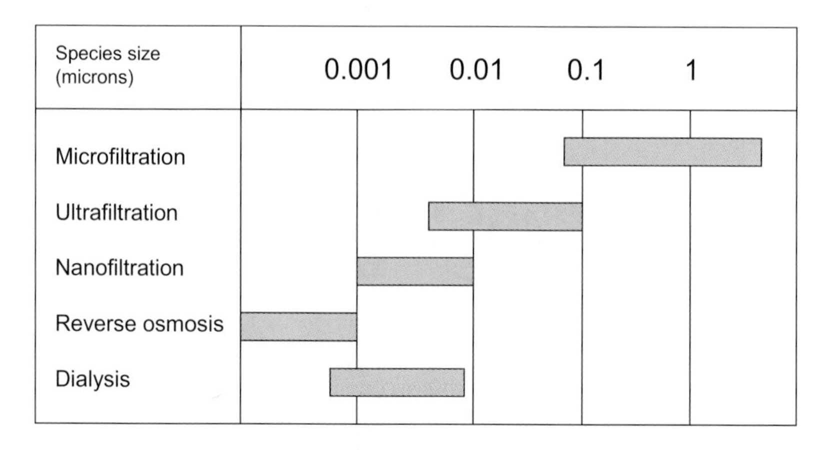

Fig. 11.11 Classification of membrane processes

Reverse osmosis (RO) membranes allow water to go through but retain all dissolved species present in the feed. In osmosis water travels from the lower solute concentration side to the higher solute concentration side of the membrane. In RO the reverse takes place due to the application of transmembrane pressure. The normal transmembrane pressure range in RO is 200 to 300 psig. Recently developed membranes allow flow of water at as low as 125 psig transmembrane pressure.

11.3. Membrane equipment

Membranes are housed within devices called membrane modules. The different types of membrane modules are:

1. Stirred cell module
2. Flat sheet tangential flow (TF) module
3. Spiral wound membrane module
4. Tubular membrane module
5. Hollow fibre membrane module

Fig. 11.12 shows a stirred cell membrane module. It is basically a stirred tank with a membrane disc fitted at the bottom. The membrane disc sits on a grooved plate which facilitates permeate collection. The stirred cell is filled with the feed and this is pressurized by compressed air or nitrogen. The stirred cell module can also be operated by continuously pumping in the feed from a reservoir. The stirrer keeps the content of the stirred tank well mixed and provides the desired shear rate on the membrane. Stirred cells are usually operated in the dead-end mode, i.e. the feed is the only thing going into the module and the permeate is the only thing going out of the module. If required, a stirred cell can be operated in a "pseudo cross-flow" mode, i.e. with continuous retentate withdrawal. Stirred cell modules are useful for small scale manufacturing and research applications. These are more commonly used for UF and MF. Stirred cell modules provide uniform conditions near the membrane surface. These are therefore very useful for process development and optimization work.

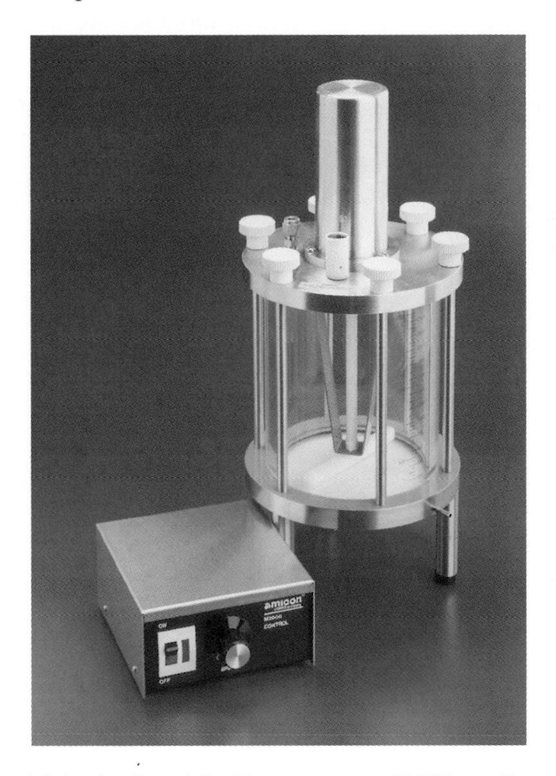

Fig. 11.12 Stirred cell module (Photo courtesy of Millipore Corporation)

The flat sheet tangential flow module has a design similar to that of a plate and frame filter press. The basic flat sheet tangential flow unit consists of a shallow channel with rectangular flat sheet membranes on one or both sides. The membranes are supported on grooved plates which allow easy collection of permeate. The feed is pumped into the channel and the permeate which crosses the membrane is collected by the grooved plates. Most flat sheet membrane modules are based on membrane cassettes (see Fig. 11.13) which contain basic flat sheet membrane units connected in series. A production-scale membrane module typically contains several of these cassettes within it connected in parallel. One of the main advantages of the flat sheet tangential flow membrane module is the ease of cleaning and replacement of defective membranes. Other advantages include ability to handle viscous feeds and feed with high levels of suspended solids. The main disadvantage is the relatively low packing density (i.e. membrane area to module volume ratio). TF membrane modules are used for UF, MF and NF.

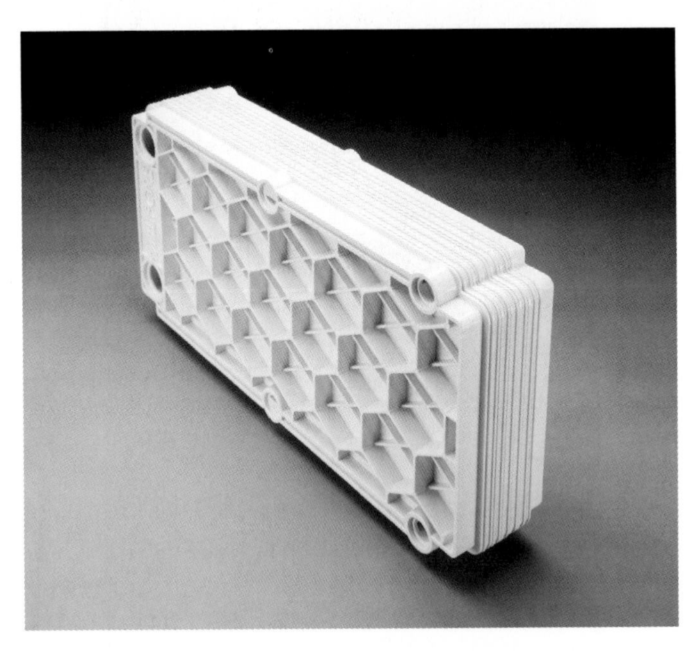

Fig. 11.13 Flat sheet TF cassette (Photo courtesy of Millipore Corporation)

The spiral wound membrane module uses flat sheet membranes. The membrane is first folded up like an envelope and this is wound up in a spiral form using feed spacers. This is then housed within a cylinder which allows the feed to be distributed and the retentate and the permeate to be collected. Fig. 11.14 shows a spiral wound membrane module. The feed is usually pumped into the space outside the envelope while the permeate collected inside the envelope runs out from the end of the module. Advantages include high membrane packing density and relatively low manufacturing cost. Disadvantages include problems with handling suspended solids and difficulty in cleaning. Spiral wound membrane modules are mainly used for RO, NF and UF.

Fig. 11.14 Spiral wound membrane module (Photo courtesy of Millipore Corporation)

A tubular membrane module is made up of several tubular membranes arranged as tubes are in a shell and tube type heat exchanger (see Fig. 11.15). The feed stream is pumped into the lumen (i.e. the inside) of the tubular membranes from one end and the retentate is collected from the other end. The permeate passes through the membrane and is collected on the shell side. Advantages include low fouling, relatively easy cleaning, easy handling of suspended solids and viscous fluids and the ability to replace or plug a damaged membrane. Disadvantages include high capital cost, low packing density, high pumping costs, and high dead volume. Tubular membranes are used for all types of pressure driven separations.

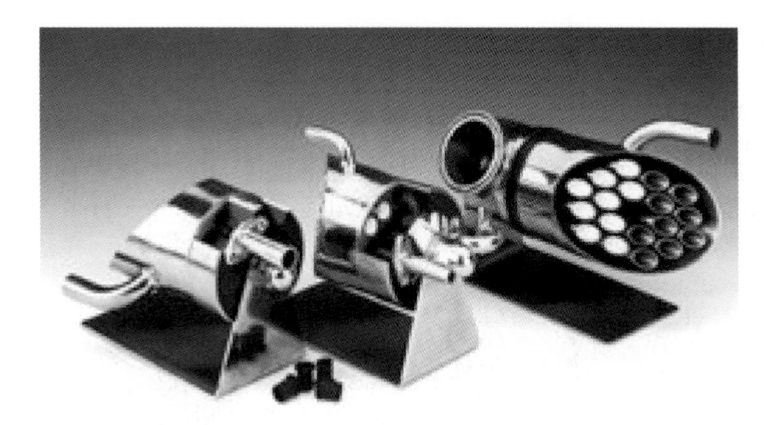

Fig. 11.15 Tubular membrane module (Photo courtesy of PCI Membranes)

The hollow fibre membrane module is similar in design to the tubular membrane module, i.e. like a shell and tube heat exchanger. Fig. 11.16 shows hollow fibre membrane modules within which large numbers of hollow fibres are potted in parallel. Advantages include low pumping cost, very high packing density, the possibility of cleaning with back-flushing, and low dead volume. Disadvantages include the fragility of the fibres, inability to handle suspended solids and need to replace entire module in case of fibre damage. Hollow fibre membrane modules are used for UF, MF and dialysis.

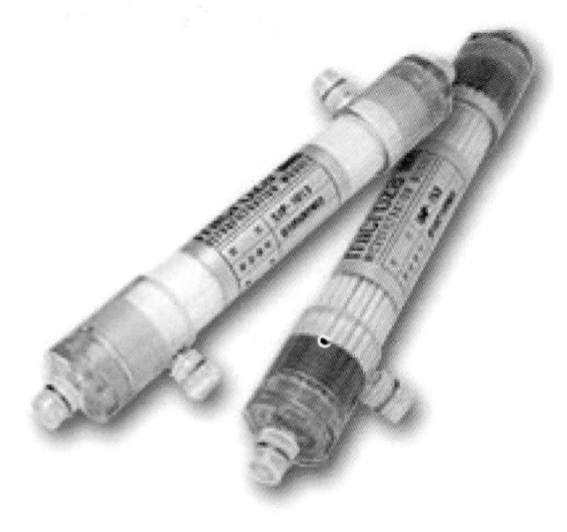

Fig. 11.16 Hollow fibre membrane module (Photo courtesy of Asahi Kasei Corporation)

11.4. Ultrafiltration

UF membranes can retain macromolecular solutes. Solute retention is mainly determined by solute size. However, other factors such as solute-solute and solute-membrane interactions can affect solute retention. Ultrafiltration is used for:

1. Concentration of solutes
2. Purification of solvents
3. Fractionation of solutes
4. Clarification

UF is attractive because of the high throughput of product, low process cost and ease of scale-up. UF is now widely used for processing therapeutic drugs, enzymes, hormones, vaccines, blood products and antibodies. The major areas of application are listed below:

1. Purification of proteins and nucleic acids
2. Concentration of macromolecules
3. Desalting, i.e. removal or salts and other low molecular weight compounds from solution of macromolecules
4. Virus removal from therapeutic products

Ultrafiltration separates solutes in the molecular weight range of 5 kDa to 500 kDa. UF membranes have pores ranging from 1 to 20 nm in diameter. Most UF membranes are anisotropic, with a thin "skin layer", typically around 10 μm thick fused on top of a microporous backing layer. The skin layer confers selectivity to the membrane while the microporous backing layer provides mechanical support. The ability of an ultrafiltration membrane to retain macromolecules is traditionally specified in terms of its molecular cut-off (MWCO). A MWCO value of 10 kDa means that the membrane can retain from a feed solution 90% of the molecules having molecular weight of 10 kDa. This is a highly subjective definition since it does not specify any other conditions such as feed concentration, transmembrane pressure and so on. However, the MWCO provides a good starting point for selecting a membrane for a given application.

The flow of a solvent through ultrafiltration membranes can be described in terms of a pore flow model which assumes ideal cylindrical pores aligned normal to the membrane surface:

$$J_v = \frac{\varepsilon_m d_p^2 \Delta P}{32 \mu l_p} \tag{11.1}$$

Where

J_v = volumetric permeate flux (m^3/m^2 s or m/s)

ε_m = membrane porosity (-)

d_p = average pore diameter (m)

ΔP = transmembrane pressure (Pa)

μ = viscosity (kg/m s)

l_p = average pore length (m)

Ultrafiltration is usually operated in the cross-flow mode where the feed flows parallel to the membranes surface as shown in Fig. 11.17. The dead ended mode of operation where the feed is pumped normal to the membrane surface is rarely used in ultrafiltration processes. The transmembrane pressure in cross-flow UF is given by:

$$\Delta P = \frac{P_i + P_o}{2} - P_p \qquad (11.2)$$

Where

P_i = inlet pressures on the feed side (Pa)

P_o = outlet pressures on the feed side (Pa)

P_p = pressure on the permeate side (Pa)

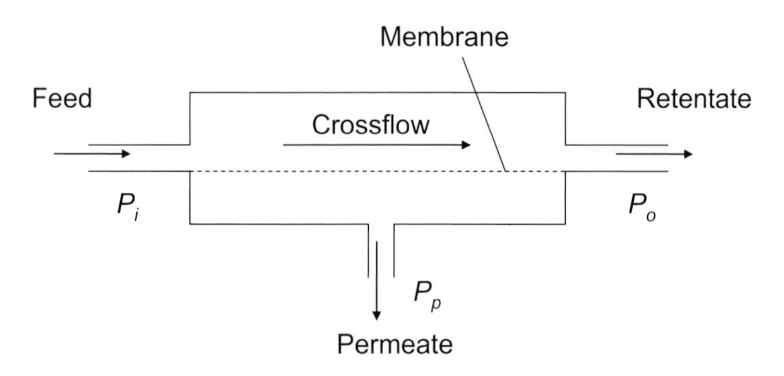

Fig. 11.17 Crossflow membrane filtration

The transmembrane pressure in UF is usually generated by pressurizing the feed side with a valve on the retentate line. The transmembrane pressure could also be generated by negative pressure, i.e. by providing a pump on the permeate side which would draw the permeate by suction. The first option is preferred in constant pressure ultrafiltration while the second option is used for constant flux ultrafiltration.

As most ultrafiltration membranes can not be visualized as having parallel cylindrical pores, a parameter, the membrane hydraulic resistance is used for calculating permeate flux:

$$J_v = \frac{\Delta P}{R_m} \tag{11.3}$$

Where

R_m = membrane hydraulic resistance (Pa s/m)

When a solution of macromolecules is ultrafiltered, the retained macromolecules accumulate near the membrane surface. This is known as concentration polarization. At steady state, a stable concentration gradient exists near the membrane owing to back diffusion of solute from the membrane surface. Concentration polarization not only offers extra hydraulic resistance to the flow of solvent but also results in the development of osmotic pressure which acts against the applied transmembrane pressure. Therefore:

$$J_v = \frac{\Delta P - \Delta \pi}{R_m + R_{cp}} \tag{11.4}$$

Where

$\Delta \pi$ = osmotic pressure (Pa)

R_{cp} = resistance due to concentration polarization layer (Pa s/m)

If the solute build–up is extensive, a gel layer may be formed on top of the membrane. In such a situation, the equation needs to be modified to account for the gel layer resistance:

$$J_v = \frac{\Delta P - \Delta \pi}{R_m + R_{cp} + R_g} \tag{11.5}$$

Where

R_g = gel layer resistance (Pa s/m)

The formation of the concentration polarization layer and the gel layer can be inferred from the permeate flux versus transmembrane pressure profile for an ultrafiltration process involving partially or totally retained solutes. Fig. 11.18 shows a typical UF flux – pressure profile. At lower values of transmembrane pressure, the permeate flux increases linearly with increase in pressure, almost coinciding with the solvent profile. However, as the pressure is further increased, there is deviation from the solvent profile, this being due to concentration polarization. At very high transmembrane pressures, the permeate flux usually plateaus off, clearly suggesting the formation of a gel layer. Beyond this point, increasing the transmembrane pressure has a negligible effect on the permeate flux, this value of permeate flux being referred to as the limiting flux (J_{lim}). The profile shown in Fig. 11.18 can be divided into two zones: the pressure dependent zone consisting of the linear part of the profile and the non-linear part due to concentration polarization in which increase in transmembrane pressure still results in some increase in permeate flux, and the pressure independent zone where the limiting flux has been reached. Ultrafiltration processes are usually operated in the pressure dependent zone.

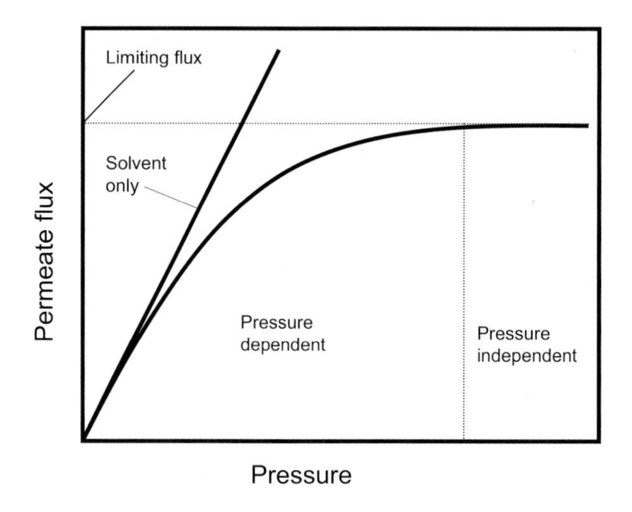

Fig. 11.18 Effect of transmembrane pressure on permeate flux

Due to the build-up of rejected solute molecules near the membrane surface and the resulting back diffusion of solutes into the feed, a concentration profile of solute molecules such as shown in Fig. 11.19 is obtained. At steady state, a material balance of solute molecules in a control volume within the concentration polarization layer yields the following differential equation:

$$J_v C - J_v C_p + D\frac{dC}{dx} = 0 \qquad (11.6)$$

Where

C = solute concentration (kg/m^3)
C_p = solute concentration in permeate (kg/m^3)
D = diffusivity (m^2/s)

Integrating equation (11.6) with boundary conditions:
$C = C_w$ at $x = 0$ and $C = C_b$ at $x = \delta_b$, we get:

$$J_v = k \ln\left(\frac{C_w - C_p}{C_b - C_p}\right) \qquad (11.7)$$

Where
C_b = bulk feed concentration (kg/m^3)
C_w = wall concentration (kg/m^3)
k = mass transfer coefficient (m/s) (= D / δ_b)
δ_b = boundary layer thickness (m)

Equation (11.7) is known as the concentration polarization equation for partially rejected solutes. For total solute rejection, i.e., when $C_p = 0$, this equation can be written as:

$$J_v = k \ln\left(\frac{C_w}{C_b}\right) \qquad (11.8)$$

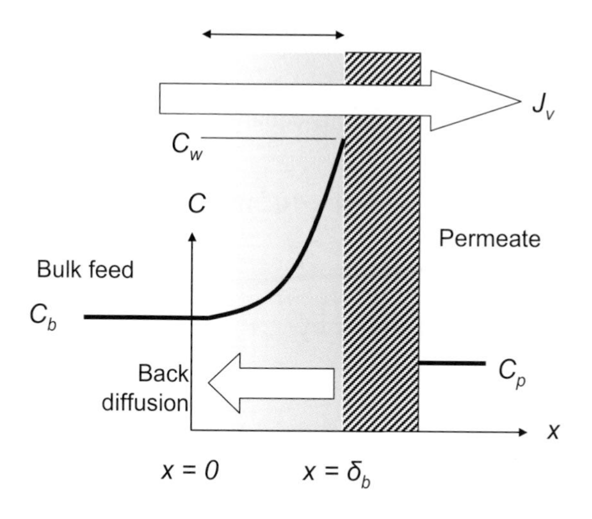

Fig. 11.19 Concentration polarization

When the solute concentration at the membrane surface reaches the gelation concentration of the macromolecule, there can be no further increase in the value of C_w. Thus:

$$J_{\lim} = k \ln\left(\frac{C_g}{C_b}\right) \tag{11.9}$$

Where

C_g = gelation concentration (kg/m^3)

Equation (11.9) is referred to as the gel polarization equation. It indicates that when C_w equals C_g the permeate flux is independent of the TMP. In the pressure independent region, the permeate flux for a given feed solution is only dependent on the mass transfer coefficient. For a particular mass transfer coefficient the plateau permeate flux value is referred to as its limiting flux (J_{lim}).

Example 11.1

A protein solution (concentration = 4.4 g/l) is being ultrafiltered using a spiral wound membrane module, which totally retains the protein. At a certain transmembrane pressure the permeate flux is 1.3×10^{-5} m/s. The diffusivity of the protein is 9.5×10^{-11} m^2/s while the wall concentration at this operating condition is estimated to be 10 g/l. Predict the thickness of the boundary layer. If the permeate flux is increased to 2.6×10^{-5} m/s while maintaining the same hydrodynamic conditions within the membrane module, what is the new wall concentration?

Solution

Where there is total solute retention, equation (11.8) can be used. This equation can be written as:

$$k = \frac{J_V}{\ln\left(\frac{C_w}{C_b}\right)} = \frac{1.3\times10^{-5}}{\ln\left(\frac{10}{4.4}\right)} \text{ m/s} = 1.584 \times 10^{-5} \text{ m/s}$$

The mass transfer coefficient is give by:

$$k = \frac{D}{\delta_b}$$

Therefore:

$$\delta_b = \frac{9.5\times10^{-11}}{1.584\times10^{-5}} \text{ m} = 5.99 \times 10^{-6} \text{ m} = 5.99 \text{ microns}$$

When J_v is increased to 2.6×10^{-5} m/s and k remains the same, the wall concentration can be obtained from the concentration polarization equation for a totally retained solute, written as shown below:

$$C_w = C_b \exp\left(\frac{J_v}{k}\right) = 4.4 \times \exp\left(\frac{2.6 \times 10^{-5}}{1.584 \times 10^{-5}}\right) \text{g/l} = 22.727 \text{ g/l}$$

Fig. 11.20 shows the effect of feed concentration on permeate flux in an ultrafiltration process. Quite clearly, at a particular transmembrane pressure, the permeate flux decreases as the feed concentration is increased. Also, the limiting flux value decreases as the feed concentration is increased. Fig. 11.21 shows plots of limiting flux versus ln (C_b) at different mass transfer coefficient values. From these plots, it may be inferred that for a given feed concentration, the limiting flux increases with increase in mass transfer coefficient. Such plots can be used to calculate C_g and k from experimental data.

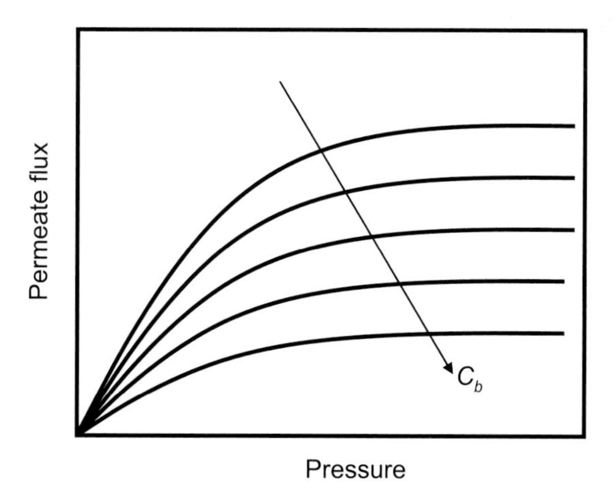

Fig. 11.20 Effect of feed concentration on permeate flux

The solute mass transfer coefficient (k) is a measure of the hydrodynamic conditions within a membrane module. The mass transfer coefficient can be estimated from correlations involving the Sherwood number (Sh), the Reynolds number (Re), and the Schmidt number (Sc):

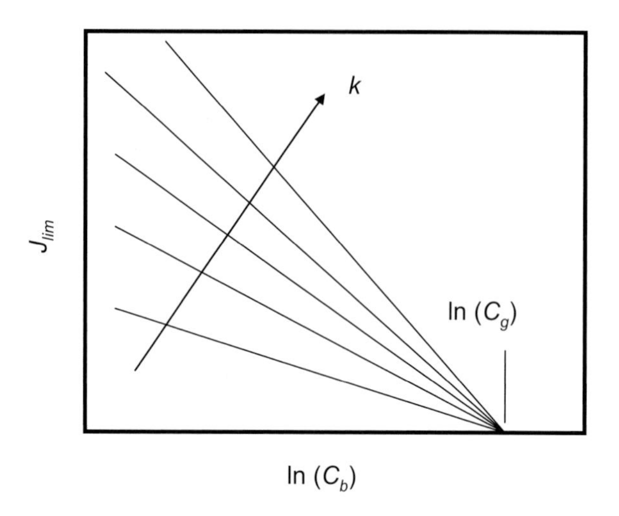

Fig. 11.21 Plots of limiting flux versus ln (C_b)

$$Sh = \left(\frac{kd}{D}\right)$$

$$Re = \left(\frac{du}{\nu}\right)$$

$$Sc = \left(\frac{\nu}{D}\right)$$

Where

d = hydrodynamic diameter of module (m)
u = average cross-flow velocity (m/s)
ν = kinematic viscosity (m^2/s)

These correlations are based on heat and mass transfer analogy. In the case of fully developed laminar flow, the Graetz-Leveque correlation can be used:

$$Sh = 1.62\, Re^{0.33} Sc^{0.33} \left(\frac{d}{l}\right)^{0.33} \quad (11.10)$$

Where

l = length of the module (m)

For turbulent flow (i.e. $Re > 2000$), the Dittus-Boelter correlation can be used:

$$Sh = 0.023\, Re^{0.8} Sc^{0.33} \tag{11.11}$$

The permeate flux in an ultrafiltration process determines its productivity. As already discussed, the permeate flux depends primarily on the properties of the membrane and the feed solution. For a particular membrane-feed system, the permeate flux depends on the transmembrane pressure and the solute mass transfer coefficient (which affects the concentration polarization). Permeate flux is also affected by membrane fouling: in constant transmembrane pressure ultrafiltration the permeate flux decreases with time due to fouling (see Fig. 11.8). Achieving a high permeate flux in an ultrafiltration process is important and this can be done by controlling the extent of concentration polarization and membrane fouling. Some ways by which permeate flux can be enhanced are listed below:

1. By increasing the cross-flow rate
2. By creating pulsatile or oscillatory flow on the feed side
3. By back flushing the membrane
4. By creating turbulence on the feed side using inserts and baffles
5. By sparging gas bubbles into the feed

Most membrane manufacturers use the MWCO (molecular weight cut-off) for describing the solute retention characteristics of ultrafiltration membranes. For instance if a membrane is rated as having 10 kDa MWCO, it is being implied that this membrane would give 90% retention of a solute having a molecular weight of 10 kDa. This definition is rather arbitrary and the MWCO is at best taken as a preliminary guideline from membrane selection. It is quite common to find that similarly rated membranes from different manufacturers have quite different solute retention behaviour. The retention of a solute by a membrane primarily depends of on the solute diameter to pore diameter ratio. It is also strongly dependent on the solute shape, solute charge, solute compressibility, solute-membrane interactions (which depend on the solution conditions) and operating conditions (such as cross-flow velocity and transmembrane pressure). If a solute is not totally retained (or rejected), the amount of solute going through the membrane can be quantified in terms of the membrane intrinsic rejection coefficient (R_i) or intrinsic sieving coefficient (S_i):

$$R_i = 1 - \frac{C_p}{C_w} = 1 - S_i \tag{11.12}$$

The solute concentration on the membrane surface (C_w), is difficult to determine using simple experimental methods. More practical parameters such as the apparent rejection coefficient (R_a) or the apparent sieving coefficient (S_a) are frequently preferred:

$$R_a = 1 - \frac{C_p}{C_b} = 1 - S_a \tag{11.13}$$

Early attempts to correlate the ratio of solute diameter and the pore diameter with the apparent rejection coefficient yielded equations of the form shown below:

$$R_a = (\lambda(2 - \lambda))^2 \text{ for } \lambda < 1 \tag{11.14}$$

$$R_a = 1 \text{ for } \lambda \geq 1 \tag{11.15}$$

Where

λ = d_i / d_p = solute-pore diameter ratio (-)

d_i = solute diameter (m)

d_p = pore diameter (m)

It is now recognized that the rejection coefficients depend not only on solute and membrane properties but also on operating and environmental parameters such as feed concentration, solution pH, ionic strength, system hydrodynamics and permeate flux. The transmission of charged solutes is particularly sensitive to pH and salt concentration. The intrinsic rejection coefficient increases with increase in permeate flux, transmembrane pressure and solute molecular weights, but is independent of the system hydrodynamics. The apparent rejection coefficient increases with increasing molecular weight, cross-flow velocity and feed concentration.

A correlation between the intrinsic sieving coefficient and the apparent sieving coefficient can be obtained from the concentration polarization equation by writing it as shown below:

$$J_v = k \ln\left(\frac{(C_w / C_p) - (C_p / C_p)}{(C_b / C_p) - (C_p / C_p)}\right) \tag{11.16}$$

Substituting and rearranging, we get:

$$\ln\left(\frac{S_a}{1-S_a}\right) = \ln\left(\frac{S_i}{1-S_i}\right) + \left(\frac{J_v}{k}\right) \tag{11.17}$$

If the intrinsic sieving coefficient could be considered a constant, equation (11.17) provides a way by which the mass transfer coefficient and the intrinsic sieving coefficient could be determined by plotting experimental data as shown in Fig. 11.22.

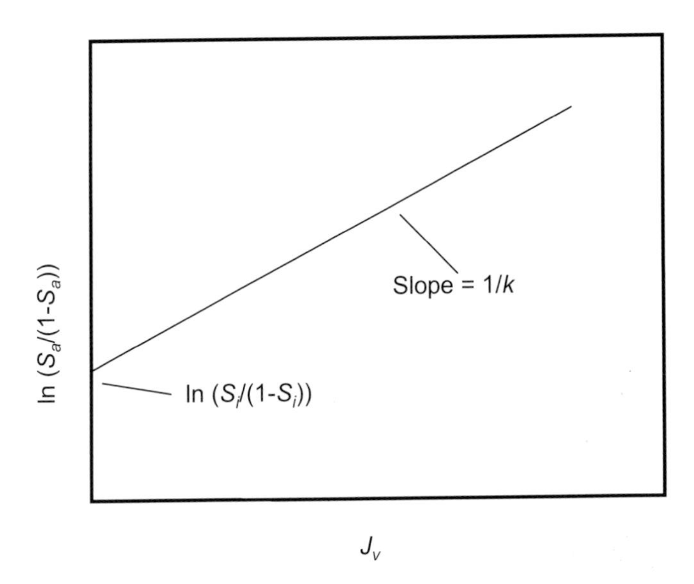

Fig. 11.22 Determination of mass transfer coefficient from sieving data

Recent research has shown that the S_i depends on J_v. However, it does not depend on the mass transfer coefficient. Fig. 11.23 shows a plot of the intrinsic sieving coefficient versus permeate flux. S_i is relative constant at very low and very high permeate flux values. There is a transition between these two constant regions in the intermediate permeate flux region. Therefore the approach for determining intrinsic sieving coefficient and mass transfer coefficient shown in Fig. 11.22 can only be used in a permeate flux range where the intrinsic sieving coefficient is constant, i.e. at very high and very low J_v.

The effect of permeate flux on the intrinsic sieving coefficient can be explained in terms of the convection-diffusion theory. According to this, the transport of a solute through an ultrafiltration membrane takes place by both diffusion and convection. When the solvent flows through the membrane at a low velocity (i.e. when J_v is low), diffusive solute transport predominates. However, when the solvent velocity is high (i.e. when J_v is high), convective solute transport predominates. Also both convective and diffusive solute transport through the membrane could be hindered due to the presence of the pore wall as well as due to pore tortuosity. All these contributing factors are combined in the equation shown below:

$$S_i = \frac{S_\infty \exp\left(\frac{S_\infty J_v \delta_m}{D_{eff}}\right)}{S_\infty + \exp\left(\frac{S_\infty J_v \delta_m}{D_{eff}}\right) - 1} \tag{11.18}$$

Where

S_∞ = asymptotic sieving coefficient (-)

δ_m = membrane thickness (m)

D_{eff} = effective diffusivity of the solute in the pore (m^2/s)

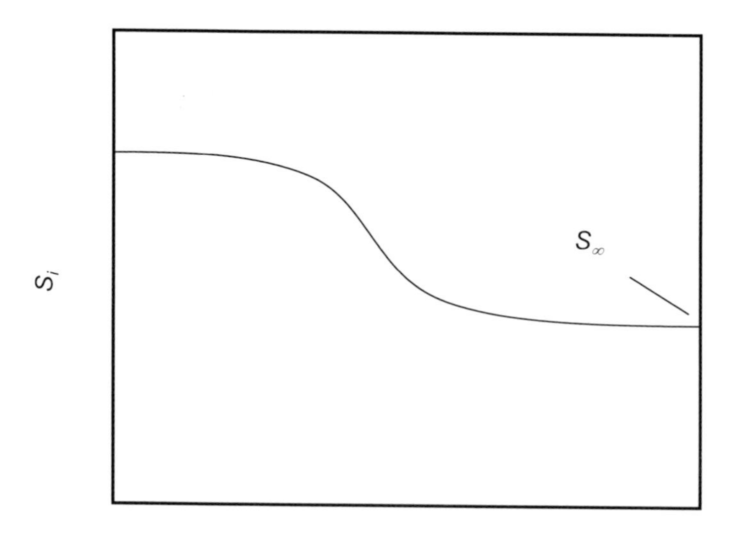

Fig. 11.23 Effect of permeate flux on intrinsic sieving coefficient

The apparent sieving coefficient depends on both the permeate flux and the mass transfer coefficient. Fig. 11.24 shows the effect of permeate flux on the apparent sieving coefficient. At low values of J_v, diffusive solute transport predominates and the apparent sieving coefficient approaches 1, i.e. the solute concentrations on both sides of the membrane are nearly the same. As J_v is increased, there is a transition from diffusive to convective solute transport. The solute transport by convective transport is hindered to a greater extent at higher solvent velocities and hence the difference in solute concentration across the membrane increases due to decreased solute transmission. This results in a decrease in the apparent sieving coefficient. However, the decrease in solute transmission results in its increased accumulation, i.e. concentration polarization. This results in an increase in the solute concentration in the permeate and is reflected in terms of an increase in the apparent sieving coefficient. From equations (11.17) and (11.18):

$$S_a = \frac{S_\infty \exp\left(\frac{S_\infty J_v \delta_m}{D_{eff}} + \frac{J_v}{k}\right)}{\left(S_\infty - 1\right)\left[1 - \exp\left(\frac{S_\infty J_v \delta_m}{D_{eff}}\right)\right] + S_\infty \exp\left(\frac{S_\infty J_v \delta_m}{D_{eff}} + \frac{J_v}{k}\right)} \tag{11.19}$$

Equation (11.19) clearly shows that the apparent sieving coefficient is affected by both permeate flux and mass transfer coefficient.

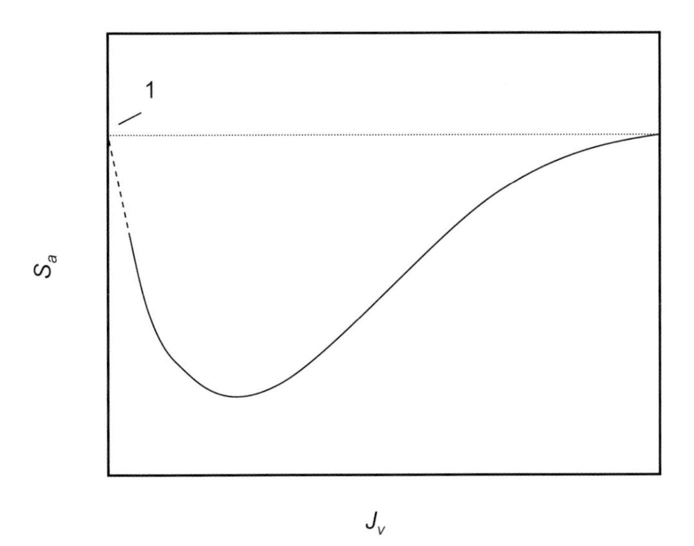

Fig. 11.24 Effect of permeate flux on apparent sieving coefficient

Example 11.2

The intrinsic and apparent rejection coefficients for a solute in an ultrafiltration process were found to be 0.95 and 0.63 respectively at a permeate flux value of 6×10^{-3} cm/s. What is the solute mass transfer coefficient?

Solution

From equation (11.17) we can write:

$$k = \frac{J_v}{\ln\left(\frac{S_a}{1-S_a}\right) - \ln\left(\frac{S_i}{1-S_1}\right)}$$

Now:

$$S_a = 1 - R_a = 1 - 0.63 = 0.37$$

$$S_i = 1 - R_i = 1 - 0.95 = 0.05$$

Therefore:

$$k = \frac{6 \times 10^{-3}}{\ln\left(\frac{0.37}{1-0.37}\right) - \ln\left(\frac{0.05}{1-0.05}\right)} \text{ cm/s} = 2.486 \times 10^{-3} \text{ cm/s}$$

Example 11.3

A solution of dextran (molecular weight = 505 kDa, feed concentration = 10 g/l) is being ultrafiltered through a 25 kDa MWCO membrane. The pure water flux values and the dextran UF permeate flux values at different TMP are given below:

ΔP (kPa)	Pure water flux (m/s)	J_v (m/s)
30	9.71E-06	6.24E-06
40	1.23E-05	7.08E-06
50	1.57E-05	7.63E-06
60	1.87E-05	8.02E-06

The osmotic pressure of dextran is given by the following correlation:

$$\log \Delta\pi = 2.48 + 1.22(C_w)^{0.35}$$

Where $\Delta\pi$ is in dynes/cm^2 and C_w is in % w/v. Calculate the membrane resistance and the mass transfer coefficient for dextran assuming that R_g and R_{cp} are negligible.

Solution

The MWCO of the membrane is 25 kDa while the molecular weight of the dextran is 505 kDa. Therefore we can safely assume that there is total dextran retention, i.e. $R_a = R_i = 1$.

Pure water ultrafiltration is governed by equation (11.3). Therefore R_m can be obtained from a plot of pure water flux versus the transmembrane pressure (the slope = $1/R_m$). From the plot:

$R_m = 3.33 \times 10^9$ Pa s/m

If resistance due to concentration polarization and gel layer are negligible, we can write equation (11.5) as shown below:

$$\Delta\pi = \Delta P - J_v R_m \qquad (11.a)$$

Using equation (11.a), the osmotic back pressure at different TMP can be calculated. The correlation for osmotic pressure given in the problem can be rearranged as shown below:

$$C_w = \left(\frac{\log_{10}(\Delta\pi) - 2.48}{1.22}\right)^{2.857} \qquad (11.b)$$

Where C_w is in % w/v and $\Delta\pi$ is in dynes/cm^2.

Using equation (11.b), the wall concentration at different TMP can be calculated. The mass transfer coefficient in an ultrafiltration process where there is total solute retention is given by:

$$k = \frac{J_v}{\ln\left(\frac{C_w}{C_b}\right)} \qquad (11.c)$$

The table shown below summarizes the results obtained from the above calculations:

ΔP (kPa)	J_v (m/s)	$\Delta\pi$ (kPa)	$\Delta\pi$ (dynes/cm^2)	C_w (%w/v)	C_w (g/l)	k (m/s)
30	6.24×10^{-6}	9.22	92208	7.63	76.31	3.07×10^{-6}
40	7.08×10^{-6}	16.42	164236	10.04	100.43	3.07×10^{-6}
50	7.63×10^{-6}	24.59	245921	11.99	119.94	3.07×10^{-6}
60	8.02×10^{-6}	33.29	332934	13.61	136.08	3.07×10^{-6}

Therefore the mass transfer coefficient is 3.07×10^{-6} m/s.

For fractionation of a binary mixture of solutes, it is desirable to achieve maximum transmission of one solute and minimum transmission of the other. Efficiency of solute fractionation is expressed in terms of the selectivity parameter:

$$\psi = \frac{S_{a1}}{S_{a2}} = \frac{(1-R_{a1})}{(1-R_{a2})} \qquad (11.20)$$

Where

Ψ = selectivity (-)
S_{a1} = apparent sieving coefficient of more transmitted solute (-)
S_{a2} = apparent sieving coefficient of less transmitted solute (-)
R_{a1} = apparent rejection coefficient of more transmitted solute (-)
R_{a2} = apparent rejection coefficient of less transmitted solute (-)

The efficiency of solute fractionation by ultrafiltration is influenced by many factors and hence very precise optimization is required. In protein-protein fractionation, the following need to be optimized:

1. Solution pH
2. Solution ionic strength
3. Cross-flow velocity
4. Transmembrane pressure or permeate flux
5. Feed concentration

Ultrafiltration processes can be operated in various different modes depending on the requirements of the process. Some of the more commonly used modes are listed below:

1. Batch concentration
2. Multi-stage continuous concentration
3. Batch diafiltration
4. Continuous diafiltration
5. Feed and bleed type operation

A concentration process refers to the selective removal of a solvent from a solution (e.g. removal of water from a protein solution). This can be achieved by using a membrane which totally retains the solute while allowing unhindered passage of solvent. A set-up used for batch concentration is shown in Fig. 11.25. A batch concentration process is usually operated at constant transmembrane pressure. Due to the continuous increase of solute concentration in the feed, the permeate flux declines with time. Batch concentration is considered to be an efficient way of processing material since the membrane has the lowest possible exposure to the feed.

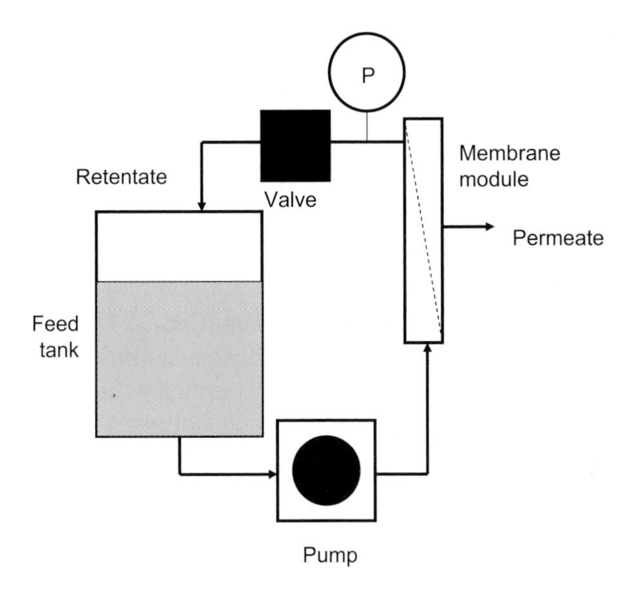

Fig. 11.25 Batch concentration

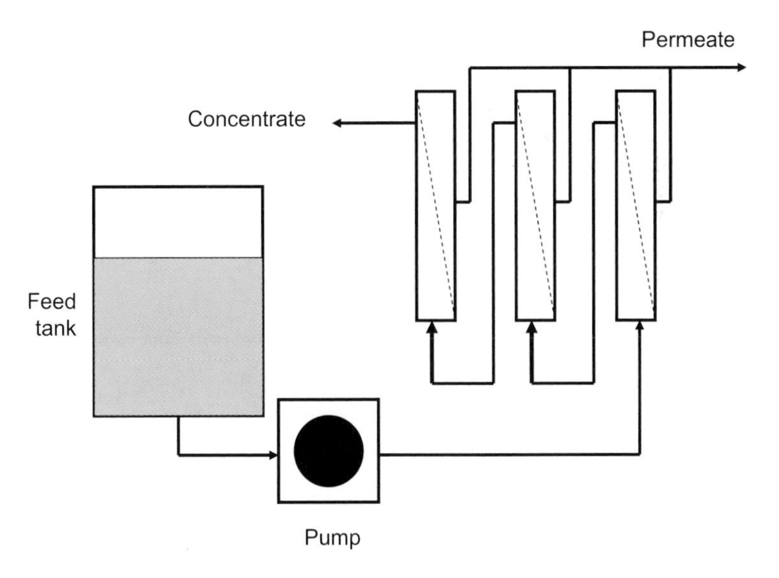

Fig. 11.26 Multi-stage continuous concentration

Concentration can be also carried out in a continuous manner but a single membrane module gives a very low concentration factor. The concentration factor is defined as the solute concentration in the product divided by that in the feed. A multi-stage process (see Fig. 11.26) is preferred for a continuous operation.

Example 11.4

A plasmid solution (concentration = 0.1 g/l) is being concentrated by ultrafiltration in a continuous manner using a using a tubular membrane module, which gives an apparent sieving coefficient of 0.02. The feed flow rate into the membrane module is 800 ml/min and the average permeate flux obtained at the operating condition is 3×10^{-5} m/s. If the membrane surface area is 0.1 m^2 predict the concentration of the plasmid in the retentate stream. Assume that the permeate flux is same at all locations on the membrane and the concentration of the plasmid increases linearly within the membrane module.

Solution

The separation process is summarized in the figure below:

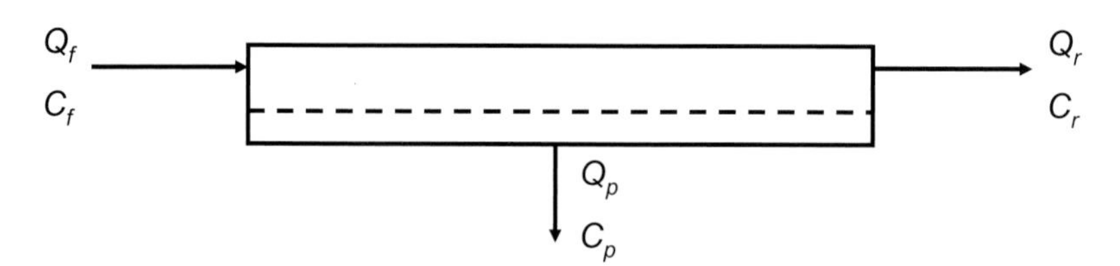

The permeate flow rate is given by the product of the volumetric permeate flux and membrane area:

$$Q_p = 3 \times 10^{-5} \times 0.1 \text{ m}^3/\text{s} = 180 \text{ ml/min}$$

Therefore the retentate flow rate is:

$$Q_r = Q_f - Q_p = 800 - 180 \text{ ml/min} = 620 \text{ ml/min}$$

The average plasmid concentration on the feed side is:

$$C_b = \frac{C_f + C_r}{2}$$

The apparent sieving coefficient is given by:

$$S_a = \frac{C_p}{C_b} = \frac{2C_p}{C_f + C_r} = 0.02 \qquad (11.d)$$

From a plasmid material balance over the membrane module we get:

$$Q_f C_f = Q_r C_r + Q_p C_p \qquad (11.e)$$

C_p and C_r are the two unknowns in equations (11.d) and (11.e). Solving these equations simultaneously we can obtain the plasmid concentration in the retentate:

$C_r = 0.1285$ g/l

Diafiltration is used for separating two solutes from one another (e.g. separation of a salt from a protein, or indeed separation of one protein from another). The membrane used should allow easy passage of the solute desired in the permeate while substantially retaining the other solute. A set-up used for batch diafiltration is shown in Fig. 11.27. The solvent lost with the permeate is replenished using fresh solvent (also called diafiltration buffer). Diafiltration can also be carried out in a continuous fashion using the set-up shown in Fig. 11.28.

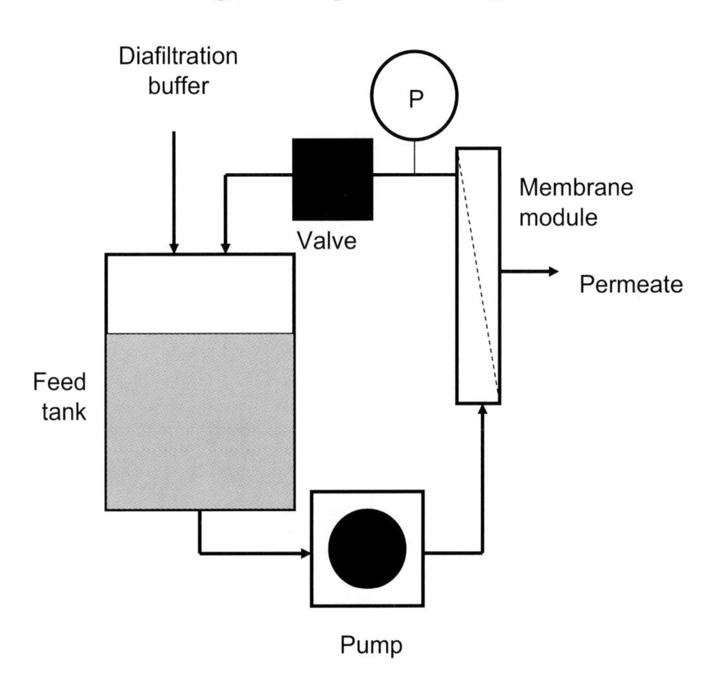

Fig. 11.27 Batch diafiltration

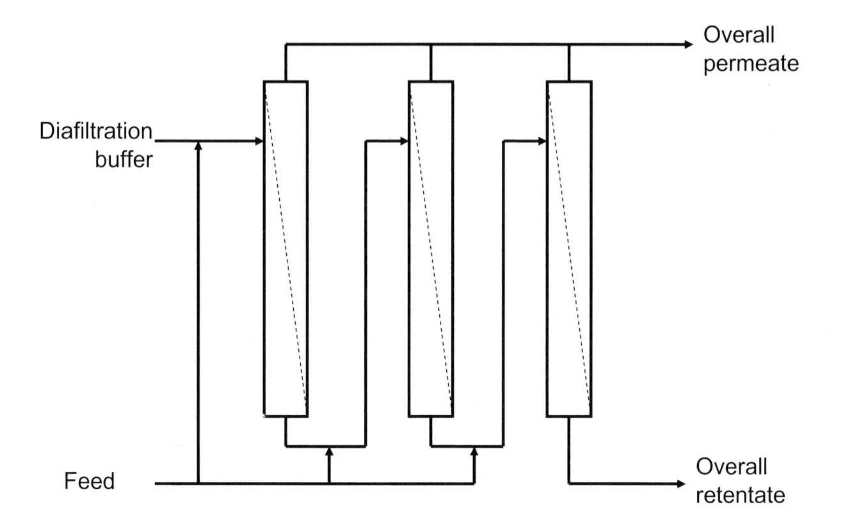

Fig. 11.28 Continuous diafiltration

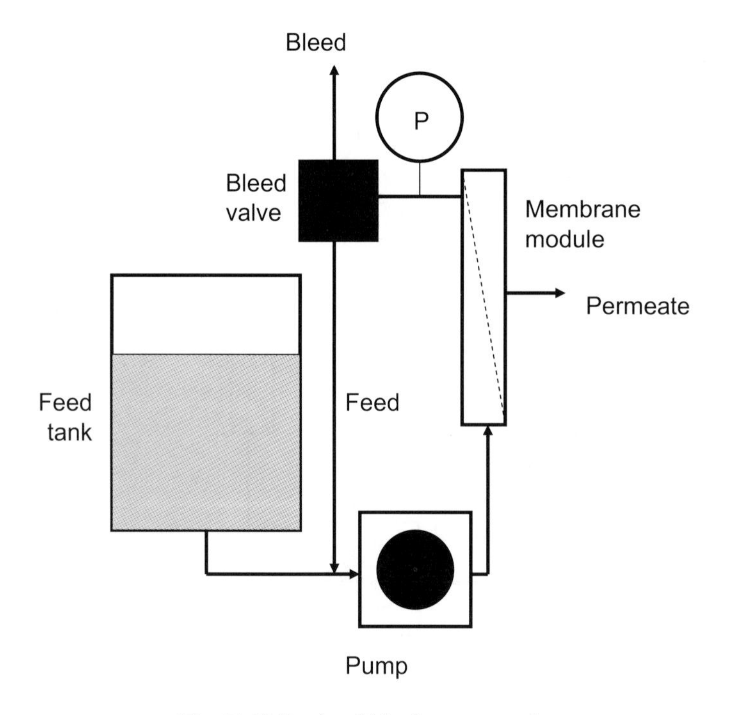

Fig. 11.29 Feed and bleed type operation

The feed and bleed type operation is used for both concentration and diafiltration. It involves the intermittent bleeding of the retentate stream as shown in Fig. 11.29. Such an operation can give very high productivity and selectivity under highly optimized conditions.

Example 11.5

A protein solution is being desalted by diafiltration in the batch mode. The membrane module has an area of 1 m^2 and the diafiltration is being carried out at a constant volumetric flux of 1×10^{-5} m/s. The volume of protein solution in the feed tank is 100 litres and the dead volume comprising the tubing, pump and membrane module is negligible. The volume of solution in the feed tank is kept constant by addition of replacement buffer (i.e. free from the salt being removed). The salt (NaCl) passes through the membrane unhindered and its initial concentration in the feed is 2 kg/m^3. If the cross-flow rate is 10 l/min calculate the time required for reducing the salt concentration to 0.2 kg/m^3.

Solution

To solve this problem we have to assume that the overall change in salt concentration on the feed side is reflected by the change in salt concentration in the feed tank. This assumption is valid only when the dead volume is negligible compared to the volume of the feed tank. The set-up described in the problem is shown below:

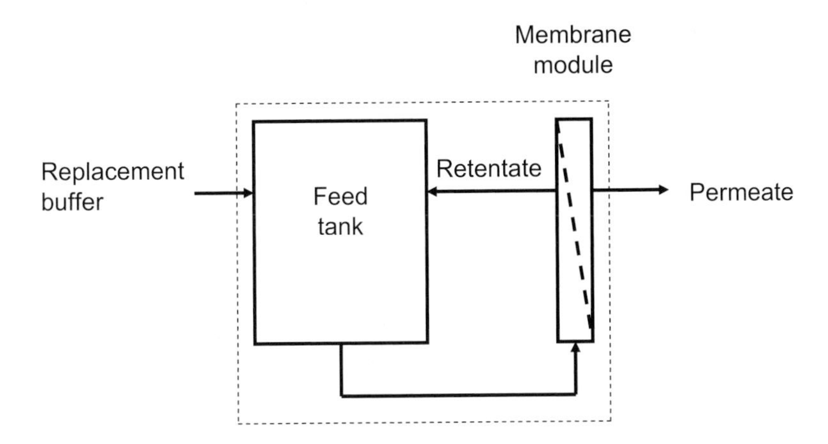

From a salt material balance over the section enclosed within the dotted lines we get:

$$0 - J_v AC = V\frac{dC}{dt} \tag{11.f}$$

Where A is the area of the membrane, C is the salt concentration in the feed tank and V is the volume of the feed tank. Rearranging equation (11.f) we get the differential equation:

$$\int dt = -\frac{V}{J_v A}\int \frac{dC}{C} \tag{11.g}$$

Integrating equation (11.g), C going from 2 kg/m^3 to 0.2 kg/m^3 and t going from 0 to t, we get:

$$t = -\frac{0.1}{1\times 10^{-5}\times 1}\times \ln\left(\frac{0.2}{2}\right) = 23025.85\ \text{s} = 383.76\ \text{min}$$

11.5. Microfiltration

Microfiltration separates micron-sized particles from fluids. The membrane modules used for microfiltration are similar in design to those used for ultrafiltration. Microfiltration membranes are microporous and retain particles by a purely sieving mechanism. Typical permeate flux values are higher than in ultrafiltration processes even though microfiltration is operated at much lower TMP. A microfiltration process can be operated either in a dead-end (normal flow) mode or cross-flow mode (see Fig. 11.30).

The various applications of microfiltration in biotechnology include:

1. Cell harvesting from bioreactors
2. Virus removal for pharmaceutical products
3. Clarification of fruit juice and beverages
4. Water purification
5. Air filtration
6. Sterilization of products

Normal flow (or dead end filtration) is used for air filtration, virus removal and sterilization processes. For most other applications, cross-flow microfiltration is preferred. Concentration polarization and membrane fouling are also observed in microfiltration. Particles have much lower diffusivity than macromolecules and consequently the extent of back diffusion of particles into the feed is negligible. Hence there is a strong tendency for cake formation. A concentration polarization scheme that is based on a stagnant film model as used in ultrafiltration cannot be applied to microfiltration since the cake layer often exceeds the boundary layer in thickness. In microfiltration, the accumulated particles are brought back to the feed solution primarily by shear induced erosion of the cake. Other factors such as particle-particle interactions and inertial lift forces can also account for back-transport of particles away from the membrane. Back flushing is a common method by which cakes formed by particles can be removed from membranes. Flux enhancement in microfiltration processes can be achieved using the same techniques as discussed in the context of ultrafiltration.

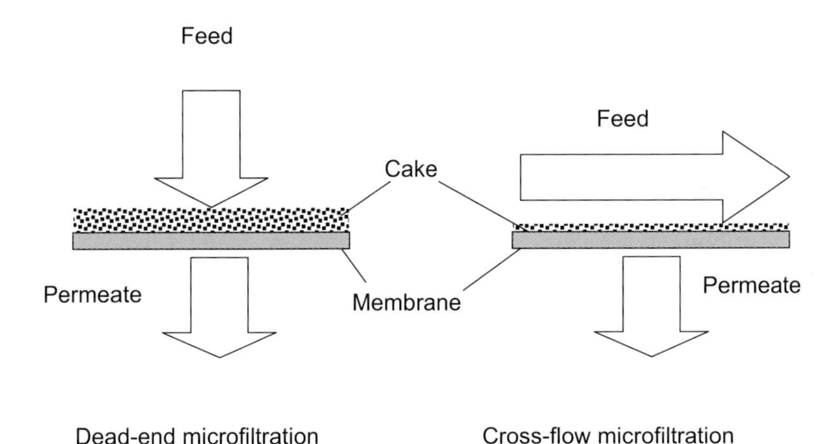

Fig. 11.30 Microfiltration

The permeate flux in microfiltration is given by:

$$J_v = \frac{\Delta P}{\mu(R_M + R_C)} \quad (11.21)$$

Where

R_M = membrane resistance (/m)

R_C = cake resistance (/m)

μ = permeate viscosity (kg/m s)

The cake resistance is given by:

$$R_C = r\frac{V_S}{A_M} \quad (11.22)$$

Where

r = Specific cake resistance (/m^2)
V_S = Volume of cake (m^3)
A_M = Area of membrane (m^2)

For micron sized particles, r is given by:

$$r = 180\left(\frac{1-\varepsilon}{\varepsilon^3}\right)\left(\frac{1}{d_s^2}\right) \tag{11.23}$$

Where
ε = Porosity of cake (-)
d_s = Mean particle diameter (m)

Example 11.6

Bacterial cells having 0.8 micron average diameter are being microfiltered in the cross-flow mode using a membrane having an area of 100 cm^2. The steady state cake layer formed on the membrane has a thickness of 10 microns and a porosity of 0.35. If the viscosity of the filtrate obtained is 1.4 centipoise, predict the volumetric permeate flux at a transmembrane pressure of 50 kPa. When pure water (viscosity = 1 centipoise) was filtered through the same membrane at the same transmembrane pressure, the permeate flux obtained was 10^{-4} m/s.

Solution

For pure water microfiltration equation (11.21) can be written as shown below:

$$J_v = \frac{\Delta P}{\mu R_M} \tag{11.h}$$

Therefore:

$$R_M = \frac{50{,}000}{(1\times 10^{-3})\times(1\times 10^{-4})} \text{ /m} = 5 \times 10^{-11} \text{ /m}$$

The specific cake resistance of the bacterial cell cake can be calculated using equation (11.23):

$$r = 180\left(\frac{1-0.35}{0.35^3}\right)\left(\frac{1}{\left(8\times 10^{-7}\right)^2}\right) \text{/m}^2 = 4.264\times 10^{15} \text{ /m}^2$$

The cake resistance can be calculated using equation (11.22):

$$R_C = r\frac{V_S}{A_M} = r\delta_C \tag{11.i}$$

In equation (11.i) δ_C is the thickness of the cake. Therefore:

$$R_C = 4.264 \times 10^{15} \times 1 \times 10^{-5} \text{ /m} = 4.264 \times 10^{10} \text{ /m}$$

The permeate flux in bacterial cell microfiltration can be obtained by equation (11.21):

$$J_v = \frac{50000}{1.4 \times 10^{-3}\left(4.264 \times 10^{10} + 5 \times 10^{11}\right)} \text{ m/s} = 6.58 \times 10^{-5} \text{ m/s}$$

11.6. Dialysis

Dialysis is a diffusion driven separation process. It is mainly used for separating macromolecules from smaller molecules. Solute separation occurs primarily because smaller solutes partition into the membrane better than bigger solutes because the degree to which the membrane restricts the entry of solutes into it increases with solute size. Smaller solutes also diffuse more rapidly than larger ones. The net results is that a dialysis membrane stops macromolecules from going through but allow smaller molecules to diffuse through.

Applications of dialysis include:

1. Removal of acid or alkali from products
2. Removal of salts and low molecular weight compounds from solutions of macromolecules
3. Concentration of macromolecules
4. Haemodialysis, i.e. purification of blood

The basic principle of dialysis is illustrated in Fig. 11.31. The concentration difference of a solute across the membrane drives its transport through the membrane. The solute flux (J) is directly proportional to the difference in solute concentration across the membrane (ΔC) and inversely proportional to the thickness of the membrane.

$$J = SD_{eff} \frac{\Delta C}{\delta_m} \tag{11.24}$$

Where

S = dimensionless solute partition coefficient (-)
D_{eff} = effective diffusivity of solute within the membrane (m^2/s)
δ_m = membrane thickness (m)

S, D_{eff} and δ_m can be combined to obtain the membrane mass transfer coefficient as shown below:

For co-current configuration:

$$\Delta C_{lm} = \frac{(C_1 - C_3) - (C_2 - C_4)}{\ln\left(\dfrac{C_1 - C_3}{C_2 - C_4}\right)} \tag{11.29}$$

Where

C_1 = feed concentration (kg/m^3)

C_2 = product concentration (kg/m^3)

C_3 = dialysing fluid concentration (kg/m^3)

C_4 = dialysate concentration (kg/m^3)

For counter-current configuration:

$$\Delta C_{lm} = \frac{(C_1 - C_4) - (C_2 - C_3)}{\ln\left(\dfrac{C_1 - C_4}{C_2 - C_3}\right)} \tag{11.30}$$

Counter-current dialysis is more commonly used. The main drawback of co-current dialysis is that, incomplete membrane utilization may take place (see Fig. 11.33). On the other hand full membrane utilization is guaranteed with counter-current dialysis. Another advantage of a counter-current dialysis is that a more uniform concentration gradient (and hence solute flux) can be maintained along the length of the membrane.

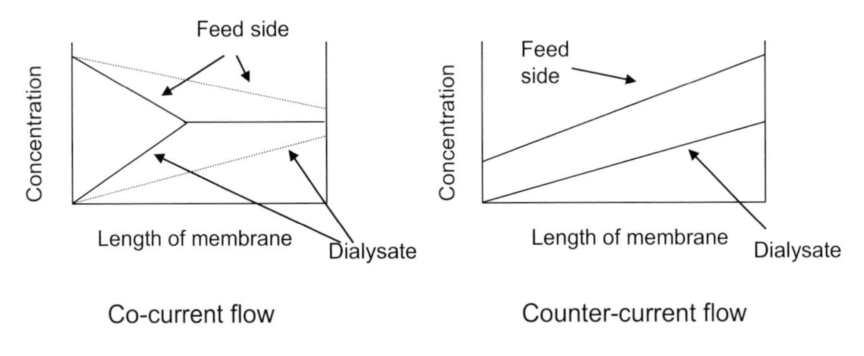

Fig. 11.33 Concentration profiles in co-current and counter-current dialysis

Example 11.7

The figure below shows a completely mixed dialyser unit. Plasma having a glutamine concentration of 2 kg/m^3 is pumped into the dialyser at a rate of 5×10^{-6} m^3/s and water at a flow rate of 9×10^{-6} m^3/s is used as the dialysing fluid. If the overall mass transfer coefficient is 2×10^{-4} m/s and the membrane area is 0.05 m^2. Calculate the steady state concentrations of glutamine in the product and dialysate streams. Assume that there is no convective transport through the membrane.

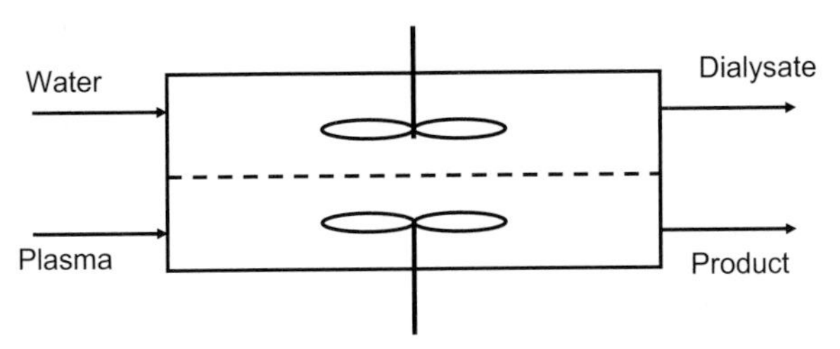

Solution

The above process is summarized below along with notations to be used in the solution:

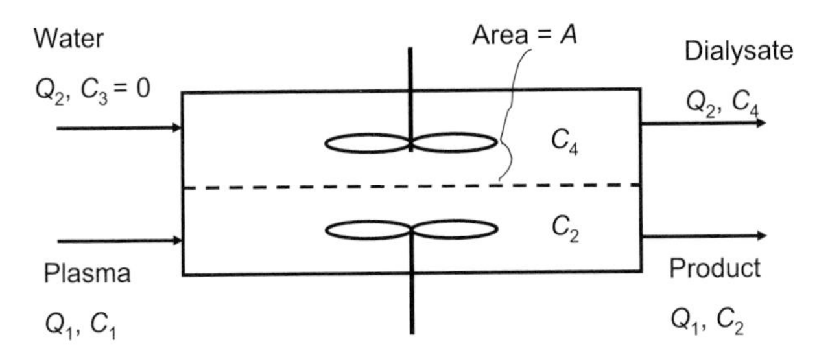

The overall material balance for glutamine gives:

$$Q_1C_1 = Q_1C_2 + Q_2C_4 \tag{11.j}$$

The glutamine flux through the membrane can be obtained using equation (11.28):

$$N = K_O(C_2 - C_4)$$

The amount of glutamine leaving with the dialysate should be equal to the product of the flux and area:

$$Q_2C_4 = K_OA(C_2 - C_4) \tag{11.k}$$

Solving equations (11.j) and (11.k) simultaneously, we get:

$C_2 = 1.027$ kg/m^3

$C_4 = 0.541\ \text{kg/m}^3$

11.7. Liquid membrane processes

Liquid membrane processes involve the transport of solutes across a thin layer of a third liquid interposed between two miscible liquids. There are two types of liquid membranes:

1. Emulsion liquid membranes (ELM)
2. Supported liquid membranes (SLM)

They are conceptually similar but different in their engineering. Emulsion liquid membranes are multiple emulsions of the water/oil/water type or the oil/water/oil type (see Fig. 11.34). The membrane phase is that interposed between the continuous (or external) phase and the encapsulated (or internal) phase. The solute mass transfer area can be dramatically increased using this membrane configuration. After transfer of solute, the membrane phase (which contains the internal phase) is separated from the external phase and then the membrane-raffinate emulsion is broken into its component phases. Fig. 11.35 shows the sequence of operations necessary for an emulsion liquid process.

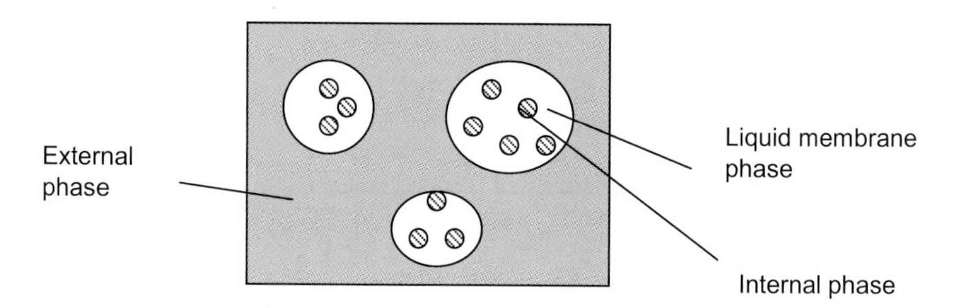

Fig. 11.34 Emulsion liquid membrane

With supported liquid membranes, the liquid membrane phase is held in place within a solid microporous inert support by capillary forces (see Fig 11.36). Very high surface areas can also be obtained using this membrane configuration. Fig. 11.37 shows solute transport in a supported liquid membrane module.

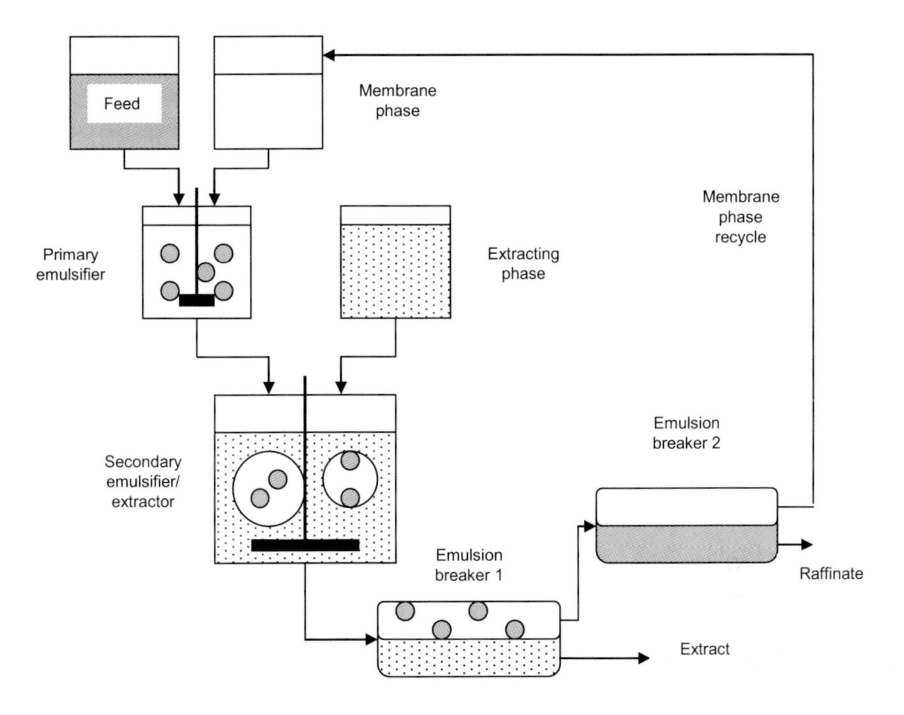

Fig. 11.35 Emulsion liquid membrane process

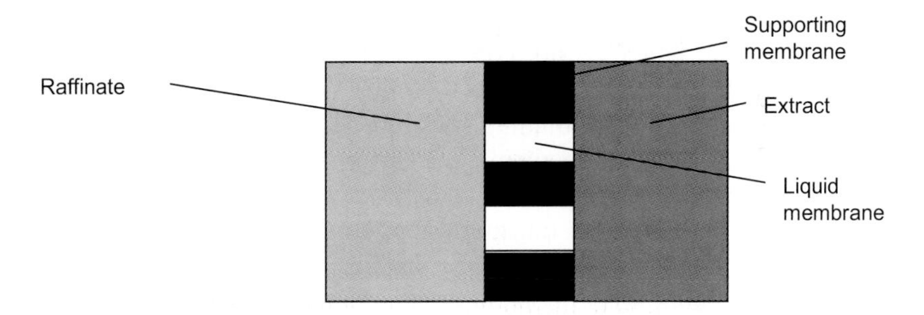

Fig. 11.36 Supported liquid membrane

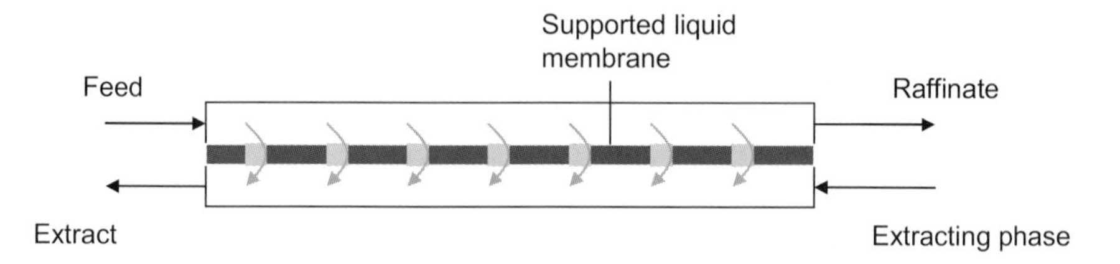

Fig. 11.37 Solute transport in supported liquid membrane module

11.8. Membrane chromatography

Adsorption and chromatographic separations are traditionally carried out using packed beds. Some of the major limitations or disadvantages of using packed beds are:

1. High pressure drop across packed beds
2. Increase in pressure drop during operation
3. Packed bed blinding by biological macromolecules
4. Dependence on intra-particle (or pore) diffusion for solute transport
5. Difficulty in scaling-up

The use of fluidised bed and expanded bed adsorption can solve some of these problems. An alternative approach has been to use stacks of synthetic microporous or macroporous membranes as adsorptive or chromatographic media. In conventional packed bed adsorption, particularly in processes using soft porous (or gel-based) media the transport of solutes to their binding sites relies heavily on diffusion and is hence slow. In contrast, in membrane adsorption, the solute transport takes place mainly by convection and hence the separation process is faster. The different solute transport steps involved in packed bed and membrane chromatography are shown in Fig. 11.38. Another important feature of membrane adsorption/chromatographic devices is the very low bed height to diameter ratio. This means that the pressure drop requirements tend to be low. Some of the more obvious advantages gained by using membranes are:

1. Low process time
2. Low process liquid requirement
3. Possibility of using very high flow rates
4. Lower pressure drop

5. Less column blinding
6. Ease of scale-up
7. Fewer problems associated with re-validation (if a disposable membrane device is used)

The first three listed advantages result from the predominance of convective transport in membrane adsorption.

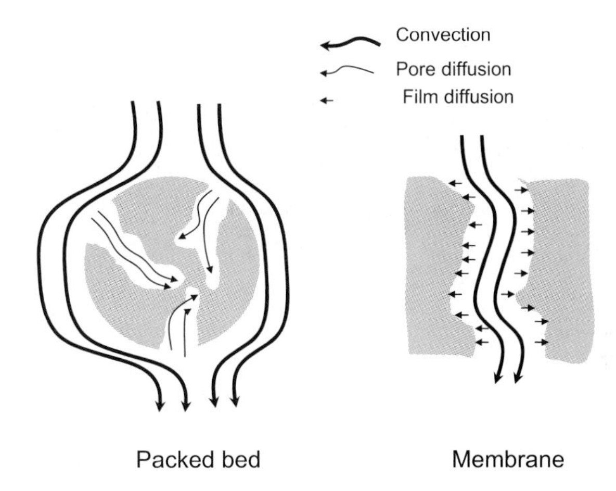

Fig. 11.38 Comparison of solute transport in packed in particulate adsorbents and in membranes

Different separation mechanisms are utilized in membrane chromatography. These include:

1. Affinity binding
2. Ion-exchange interaction
3. Hydrophobic interaction

Size exclusion based separation using membrane beds has not yet been feasible. Membrane adsorption processes are carried out in two different pulse and step input modes. The pulse input mode is similar to pulse chromatography using packed beds while the step input mode is similar to conventional adsorption. Based on the membrane geometry, three types of membrane adsorbers are used: flat sheet, radial flow and hollow fibre. Flat sheet type membrane adsorbers resemble syringe type filters commonly used for laboratory scale microfiltration. A stack of disc membranes is housed within a flat sheet type membrane module. Fig. 11.39 shows a commercial flat sheet membrane adsorber. Such adsorbers are usually operated in the bind and elute mode (i.e. the step input mode). Fig. 11.40 shows the sequence of steps along with the resultant chromatogram obtained from bind and elute chromatography. Fig. 11.41 shows a radial type membrane module in which the feed is pumped into the central core from where it is distributed into the membrane in a radially outward direction. Hollow fibre membranes are used in adsorptive separation processes by blocking off the retentate line and forcing the feed through the membrane onto the permeate side (see Fig. 11.42).

Fig. 11.39 Flat sheet type membrane adsorber (Photo copyright Sartorius AG)

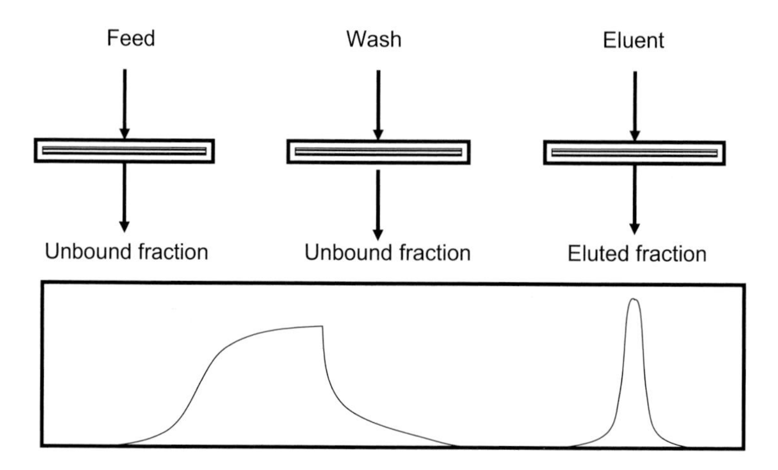

Fig. 11.40 Operation of flat sheet type membrane adsorber

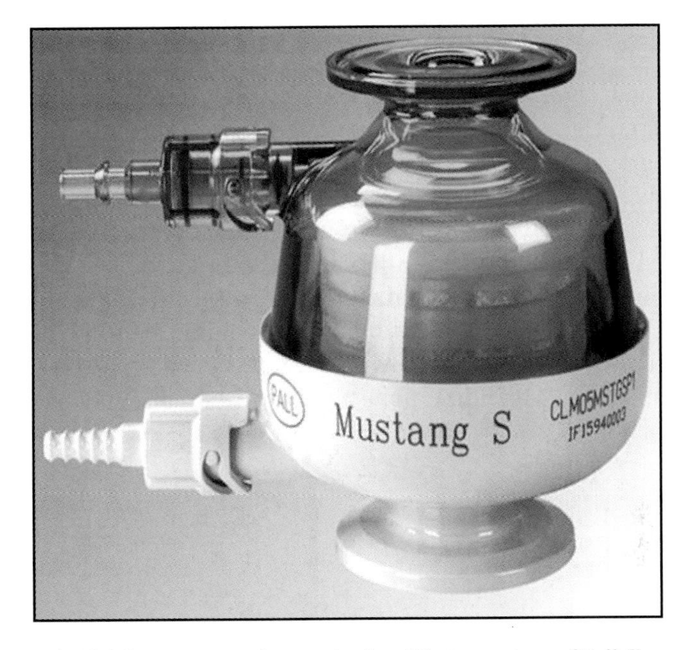

Fig. 11.41 Radial flow type membrane adsorber (Photo courtesy of Pall Corporation)

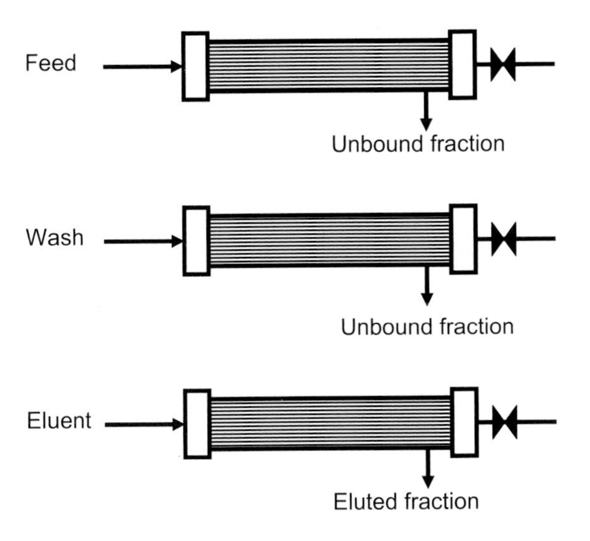

Fig. 11.42 Hollow fibre type membrane adsorber

The main limitation of membrane adsorbers is the relatively low solute binding capacity. On account of the low bed height, the plate concept and similar design concepts widely used in packed bed chromatography cannot be used satisfactorily in membrane chromatography. In a membrane adsorption process the solute molecules should have sufficient time to reach their binding sites on the pore wall during their convective transport through the membrane. A simple design calculation for membrane adsorbers is based on defining a characteristic convection time and a characteristic diffusion time (see Fig. 11.43). The diffusion time is the time taken by a solute molecule to travel from the pore centreline to the pore wall while the convection time is that taken by a solute molecule travelling along the pore centreline to travel through the membrane.

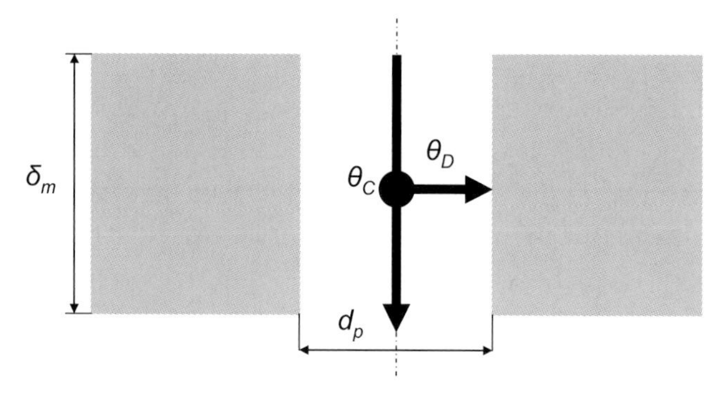

Fig. 11.43 Diffusion and convection times in membrane adsorption

The Reynolds number of the fluid flowing through the membrane pores is given by:

$$\mathrm{Re}_{pore} = \left(\frac{u_s d_{pore}}{\nu \varepsilon} \right) \tag{11.31}$$

Where

u_s = superficial velocity (m/s)

d_{pore} = average pore diameter (m)

ν = kinematic viscosity (m^2/s)

ε = porosity (-)

Generally the flow through the pores of an adsorptive membrane is laminar on account of the small value of d_{pore}. For laminar flow the centreline velocity is twice the average velocity. Therefore the residence time of a solute moving along the centreline of the fluid flowing through a membrane pore is given by:

$$\theta_C = \frac{\delta_m \tau \varepsilon}{2u_s} \tag{11.32}$$

Where

δ_m = thickness of membrane (m)

τ = pore tortuosity (-)

The time taken for a solute molecule to diffuse from the centre-line to the pore wall is given by:

$$\theta_D = \left(\frac{d_{pore}^2}{4D}\right) \tag{11.33}$$

Where

D = diffusivity (m^2/s)

In membrane chromatography θ_D should always be smaller than θ_C.

Example 11.8

An adsorptive membrane has a thickness of 2 mm and a diameter of 5 cm. The porosity of this membrane is 0.75 and the tortuosity is 1.5 while the diameter of the pores was estimated to be 2×10^{-6} m. If we are to use this membrane for adsorption of a DNA fragment (diffusivity = 9.5×10^{-12} m^2/s) from an aqueous solution, what is the maximum solution flow rate that can be used? Assume that the flow through the pores is laminar.

Solution

The area of the membrane is 1.964×10^{-3} m^2

The superficial velocity is given by:

$$u_s = \frac{Q}{1.964 \times 10^{-3}} \text{ m/s} \tag{11.l}$$

Where Q is the solution flow rate in m^3/s

The convection time can be obtained using equations (11.32) and (11.*l*):

$$\theta_C = \frac{\delta_m \tau \varepsilon}{2u_s} = \frac{2.209 \times 10^{-6}}{Q} \text{ s} \tag{11.m}$$

The diffusion time can be obtained using equation (11.33):

$$\theta_D = \frac{\left(2\times10^{-6}\right)^2}{4\times9.5\times10^{-12}}\ \text{s} = 0.1053\ \text{s}$$

For total capture of DNA fragments:

$$\theta_C > \theta_D$$

Therefore:

$$\frac{2.209\times10^{-6}}{Q} > 0.1053$$

Therefore the solution flow rate has to be lower than 2.098×10^{-5} m^3/s.

Exercise problems

11.1. A protein solution is being desalted by diafiltration using a 500 ml stirred cell ultrafiltration module, which can fit a disc membrane having 10 cm diameter. The protein solution (containing the salt) is introduced into the stirred cell and water at a constant flow rate is pumped into the module to remove the salt. The membrane totally retains the protein while allowing the salt to pass through unhindered. The effective initial concentration of salt within the module is 1 g/l. If a constant flux of 10^{-5} m/s is maintained, calculate the time required for reducing the salt concentration to 0.01 g/l.

11.2. An antibody solution is being ultrafiltered, the TMP, permeate flux and apparent sieving coefficient data being shown below. Assume that the intrinsic sieving coefficient is constant for the given flux range.

TMP (kPa)	Permeate flux (m/s)	S_a
5	9.78E-07	0.236
7.5	1.42E-06	0.381
10	1.62E-06	0.446
12.5	1.87E-06	0.521
15	2.06E-06	0.595
17.5	2.39E-06	0.664
20	2.66E-06	0.72
25	3.14E-06	0.797

Determine the intrinsic sieving coefficient and the mass transfer coefficient of the antibody. If we quadruple the mass transfer coefficient, what would the apparent sieving coefficient be at a permeate flux value of 2.06×10^{-6} m/s (assuming that the intrinsic sieving coefficient is still constant)?

11.3. The concentration of glutamine in blood is being reduced using a hollow fibre dialyser membrane with a counter-current flow arrangement. The concentration of glutamine in the inlet stream is 1 g/l while the blood flow rate and the dialysate (water) flow rate are 0.8 l/min and 0.9 l/min respectively. The surface area of the membrane is 500 cm^2 and the membrane mass transfer coefficient is 2.86×10^{-2} cm/s. The blood side mass transfer coefficient is 8.77×10^{-2} cm/s while the dialysate side mass transfer coefficient is 3.44×10^{-2} cm/s. Find the residual concentration of glutamine in blood if we assume that there is no flow of liquid across the membrane.

11.4. In an ultrafiltration process carried out at a volumetric permeate flux (J_v) of 8×10^{-6} m/s and solute mass transfer coefficient (k) of 5×10^{-6} m/s, the bulk feed and permeate concentrations of the solute were found to be 1 kg/m^3 and 0.1 kg/m^3 respectively. Predict the wall concentration of the solute. Also predict the solute concentration at the midpoint of the boundary layer.

11.5. An aqueous solution of polyethylene glycol (PEG) is being concentrated by ultrafiltration using a tubular membrane which totally retains the macromolecule but allows everything else through. The bulk feed concentration of PEG is 10 kg/m^3. At a cross-flow velocity of 0.2 m/s the permeate flux and the wall PEG concentration were found to be 5×10^{-6} m/s and 85 kg/m^3 respectively. Predict the wall PEG concentration when the permeate flux is 5×10^{-6} m/s and the cross-flow velocity is 0.1 m/s. The mass transfer coefficient for PEG can be obtained using the Porter correlation.

11.6. Two proteins A and B are being fractionated from an aqueous feed solution by ultrafiltration. At the cross-flow velocity being used the mass transfer coefficient of A is 4×10^{-6} m/s while that of B is

8×10^{-6} m/s. The fractionation is being carried out at a constant permeate flux of 1×10^{-6} m/s. The permeate flux and mass transfer coefficients combination is known to result in intrinsic sieving coefficients of 0.1 and 0.5 for A and B respectively. If we assume that each protein behaves independently of the other, predict the selectivity of separation. If the concentrations of A and B in the feed solution are 1 kg/m^3 and 3 kg/m^3 respectively, predict the composition of the permeate.

References

M. Cheryan, Ultrafiltration Handbook, Technomic, Lancaster (1986).

A.G. Fane, Ultrafiltration: Factors Influencing Flux and rejection, In: R.J. Wakeman (Ed.), Elsevier Scientific Publishing Co., Amsterdam (1986).

E. Flaschel, C. Wandrey, M.R. Kula, Ultrafiltration for Separation of Biocatalysts, in: A. Fiechter, Advances in Biochemical Engineering / Biotechnology vol. 26, Springer-Verlag, Berlin (1983).

R. Ghosh, Protein Bioseparation Using Ultrafiltration: Theory Applications and New Developments, Imperial College Press, London (2003).

E. Klein, Affinity Membranes: Their Chemistry and Performance in Adsorptive Separation Processes, Wiley Interscience, New York (1991).

M.R. Ladisch, Bioseparations Engineering: Principles, Practice and Economics, John Wiley and Sons, New York (2001).

A.S. Michaels, Ultrafiltration, in: E.S. Perry (Ed.), Progress in Separation and Purification vol. 1, Wiley Interscience, New York (1968).

W.S. Winston Ho, K.K. Sirkar, Membrane Handbook, Van Nostrand Reinhold, New York (1992).

L.Z. Zeman, A.L. Zydney, Microfiltration and Ultrafiltration: Principles and Applications, Marcel Dekker, New York (1996).

Chapter 12

Miscellaneous bioseparation processes

12.1. Introduction

In this chapter we will discuss bioseparations processes some of which have only recently been developed and others, whose use in large-scale bioprocessing is not yet widespread. Some of these could be described as hybrid bioseparation techniques since they involve combination of more than one major separation principle.

12.2. Electrophoresis

Electrophoresis refers to separation of charged solutes based on their electrophoretic mobility i.e. movement of charged molecules in response to an electric field. Electrophoresis gives excellent solute separation and this technique is extensively used for protein and nucleic acid purification and analysis. However, it use for preparative scale bioseparation is not yet extensive.

Electrophoresis is carried out by adding the mixture of solutes to be separated into a medium that conducts electricity followed by the application of an electric field across the medium (see Fig. 12.1). As a result of this, positively charged solutes migrate towards the negative electrode while the negatively charged solutes migrate towards the positive electrode. Neutral solutes will not move due to the applied electric field but they can diffuse due to concentration gradient. Depending on the type of medium in which an electrophoretic separation is carried out, electrophoresis can be classified into two types:

1. Gel electrophoresis
2. Liquid phase (or free flow) electrophoresis

Typical gel materials used for electrophoretic separation include agarose, starch and polyacrylamide. These substances form gels with high water contents and allow passage of large solutes through their porous structure. Fig. 12.2 shows the working principle gel electrophoresis. In gel electrophoresis, the molecules get separated into bands and remain within the gel after separation. These have to be subsequently extracted out from their respective locations in the gel. Proteins are usually separated by polyacrylamide gel electrophoresis (or PAGE) using vertically oriented gels. The charge on proteins depend on the pH of the medium, i.e. these will be neutral at their respective isoelectric points, positive below it and negative above it. Electrophoretic separation of proteins based on their intrinsic charge is referred to as native PAGE. In native gel electrophoresis using a set-up as shown in Fig. 12.2, all protein in the sample being separated should be negatively charged. This is done by maintaining the pH of the buffers and gel medium above the isoelectric pH of all the proteins. Where this is not possible, the pH is maintained below the isoelectric points of all the proteins in the sample and the electrodes are swapped around. This is referred to reversed polarity native PAGE. All proteins in a sample can be imparted with negative charge by boiling a protein sample in sodium dodecyl sulfate (or SDS). When proteins are boiled in SDS, they first uncoil and then coil up around SDS molecules, thus imbibing a uniform negative charge per unit length of protein. When such a sample is loaded on a gel in a set-up as shown in Fig. 12.2, all proteins move towards the positive electrode. Such a process is referred to as SDS-PAGE.

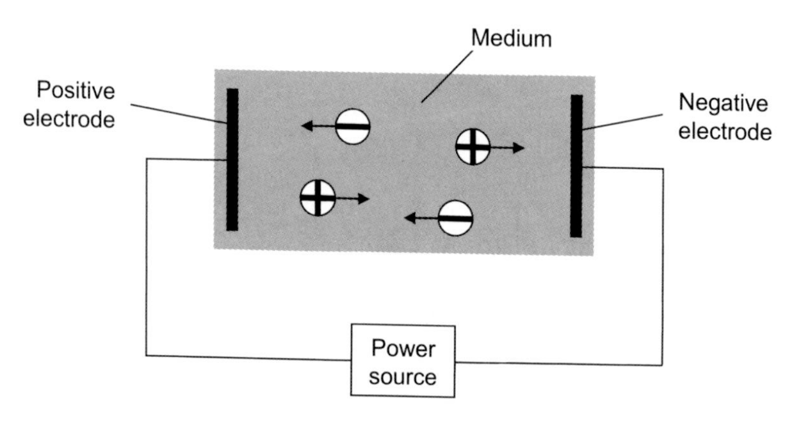

Fig. 12.1 Principle of electrophoresis

The velocity at which a charged molecule migrates in an electric field is given by:

$$v = UE \tag{12.1}$$

Where

v = velocity (m/s)
U = Electrophoretic mobility (m^2/V s)
E = Strength of electric field (V/m)

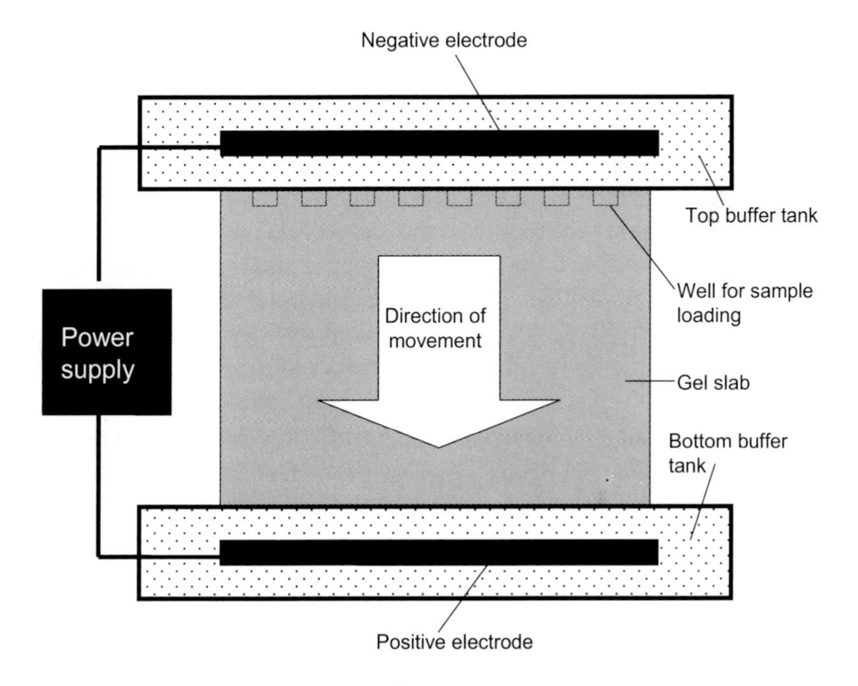

Fig. 12.2 Gel electrophoresis

Isoelectric focusing (IEF) is a specialized type of electrophoretic separation in which a pH gradient is maintained in the separation medium. A particular protein will move in such a medium only as long as it carries some net electrostatic charge. As soon as a protein reaches a zone having a pH corresponding to its isoelectric point, it does not move any further. It is easier to carry out isoelectric focusing in a gel media than in a liquid medium. 2-D (or 2 dimensional) electrophoresis is a powerful analytical separation technique, mainly used for analyzing complex protein mixtures. This technique involves the separation of the protein mixture in one direction employing IEF and in a direction perpendicular to this using SDS-PAGE. 2-D electrophoresis is widely used for proteomic analysis.

Nucleic acids are usually separated using agarose gel electrophoresis. These are negatively charged at physiological pH. Horizontal gels are more commonly used for electrophoresis of nucleic acids.

Gel electrophoresis is not particularly well suited for preparative separation. Preparative purifications can be done by liquid phase (or free-flow) electrophoresis where the separation is carried out in solutions of electrolytes, typically using a flowing liquid. The liquid flows in a shallow rectangular channel in a streamline manner (i.e. flow is laminar) and the mixture of solutes to be separate is introduced in a continuous manner or as a pulse at the channel inlet. An electric field is applied across the breadth of the flow channel. The solutes undergo motion in two-dimensions, along the length of the channel due to the flowing liquid and in a transverse direction due to the electric field. As a result of this they are segregated into the different streamlines of the flow field. The separated solutes are then flushed out of the device by the flowing liquid and collected as separate streamline fractions. Fig. 12.3 shows the principle of liquid phase electrophoretic separation. The membrane shown in the figure (the dotted lines) prevents the separated molecules from reaching the electrodes.

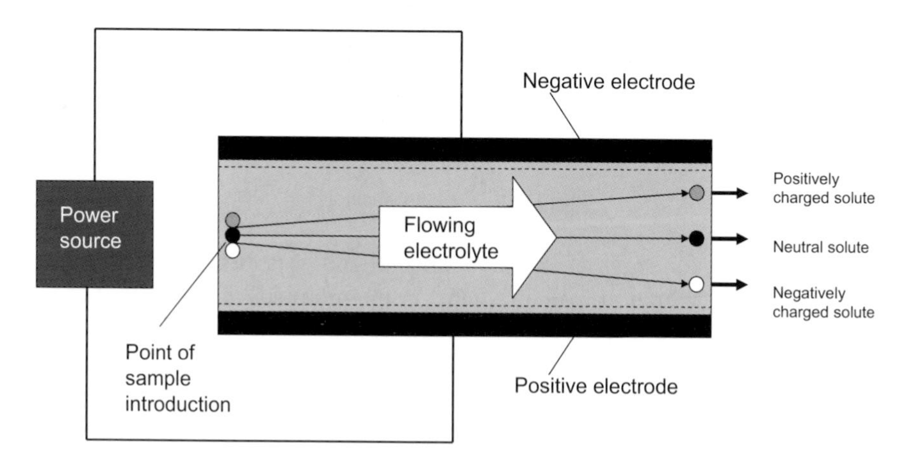

Fig. 12.3 Liquid phase (or free-flow) electrophoresis

12.3. Affinity ultrafiltration

Affinity chromatography is widely used for biological separations. One of the main problems associated with column chromatography is the predominance of diffusion in the adsorption and elution steps. Other problems include steric hindrance in confined spaces around the binding sites and the difficulty in scaling up. These problems could be addressed in different ways, one way being to carry out the affinity separation in an ultrafiltration device. The resultant separation process is called affinity ultrafiltration. The principle of affinity ultrafiltration is shown in Fig. 12.4. In affinity ultrafiltration, the affinity ligand is usually macromolecular in nature, i.e. can remain in solution but can be retained by an appropriate ultrafiltration membrane. The ultrafiltration membrane is selected such that it totally retains the macro-ligand while allowing easy passage of both the target solute and the impurities. In the binding step, the mixture to be separated is introduced into the membrane module either as a pulse or in the form of a step. The unbound impurities are then washed out of the system. Finally the target solute is eluted by using appropriate means.

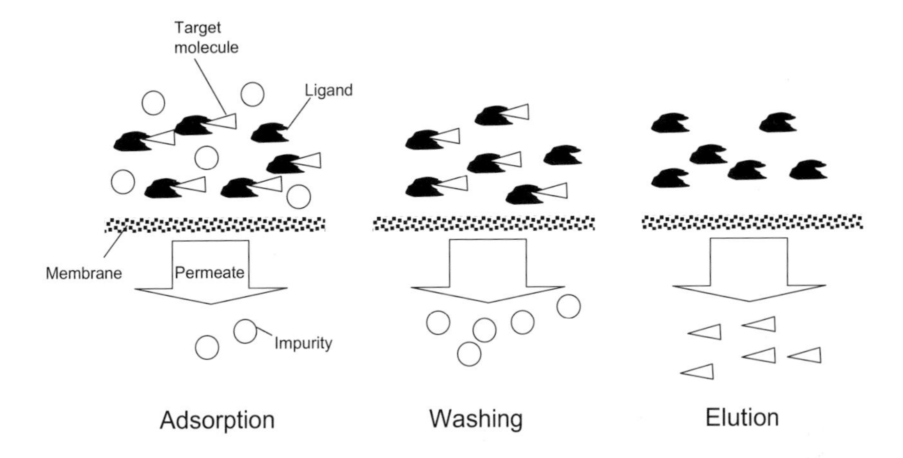

Fig. 12.4 Affinity ultrafiltration

12.4. Field-flow fractionation

Field-flow fractionation (or FFF) is mainly used to separate particles of different sizes. It can also be used to fractionate very large macromolecules and aggregated forms of macromolecules. The basic principle of FFF is shown in Fig. 12.5. A liquid is allowed to flow in a streamline (or laminar) fashion in a shallow channel and the mixture to be separated is introduced into the liquid flowing into the channel. The substances being separated are then subjected to a separation field (e.g. gravitational, electric, steric, magnetic, permeation) as a result of which the different species position themselves at different locations in the flow-field within the channel. Depending on the fluid velocity at their respective locations, the different species travel through the channel at different speeds. For instance, a substance positioned close to the lower wall of the channel will move slower than a substance positioned near the centre of the channel.

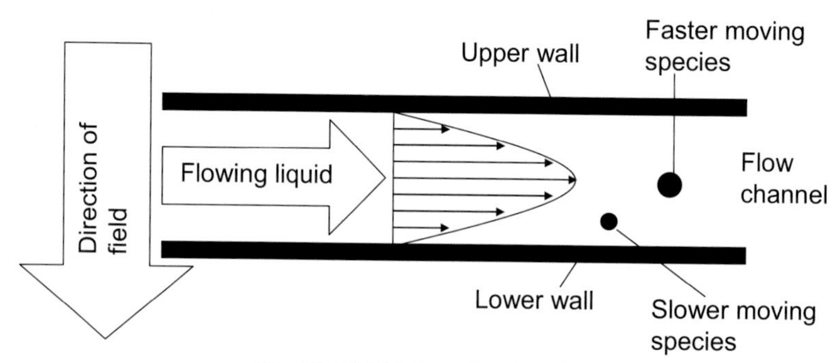

Fig. 12.5 Field-flow fractionation

References

M.R. Ladisch, Bioseparations Engineering: Principles, Practice and Economics, John Wiley and Sons, New York (2001).

P. Todd, S.R. Rudge, D.P. Petrides, R.G. Harrison, Bioseparations Science and Engineering, Oxford University Press, Oxford (2002).

M.E. Schimpf, K. Caldwell, J.C. Giddings, Field-Flow Fractionation, Wiley Interscience, New York (2000).

R. Westermeier, Electrophoresis in Practice: A Guide to Methods and Applications of DNA and Protein Separations, 4th edition, John Wiley and Sons, New York (2005).

B. Mattiasson, R. Kaul, Affinity Ultrafiltration for Protein Purification, in: T. Ngo (Ed.), Molecular Interactions in Bioseparations, Plenum Press, New York (1993).

Index

About the Author

Raja Ghosh, D.Phil. (Oxon.) holds the prestigious Canada Research Chair in Bioseparations Engineering at McMaster University, Canada. Prior to this he was a faculty member at the University of Oxford, UK. His research interests include membrane based bioseparations engineering, particularly ultrafiltration, membrane chromatography, hybrid bioseparations and membrane based bioreactors. He holds a doctorate degree in Engineering Science from the University of Oxford. His first book *Protein Bioseparation Using Ultrafiltration: Theory, Applications and New Developments* was published by Imperial College Press in 2003.